DATE DUE

Chemical and Biological Warfare

Chemical and Biological Warfare
A COMPREHENSIVE SURVEY FOR THE CONCERNED CITIZEN

ERIC CRODDY

WITH CLARISA PEREZ-ARMENDARIZ AND JOHN HART

C

Copernicus Books
An Imprint of Springer-Verlag

Published in the United States by Copernicus Books,
an imprint of Springer-Verlag New York, Inc.
A member of BertelsmannSpringer Science+Business Media GmbH

Copernicus Books
37 East 7th Street
New York, NY 10003
www.copernicusbooks.com

Library of Congress Cataloging-in-Publication Data
Croddy, Eric
    Chemical and biological warfare : a comprehensive survey for the concerned citizen / Eric
Croddy with Clarisa Perez-Armendariz and John Hart.
        p.   cm.
    Includes bibliographical references and index.
    ISBN 0-387-95076-1 (alk. paper)
    1. Chemical warfare.   2. Biological warfare   I. Title.
    UG447.C755   2001
    355'.34—dc21   2001054929

Manufactured in the United States of America.
Printed on acid-free paper.

9 8 7 6 5 4 3 2 1

ISBN 0-387-95076-1   SPIN 10771289

Contents in Brief

PART ONE: GAS, BUGS, AND COMMON SENSE   I

    Chapter 1:   The Fog of War   3
    Chapter 2:   Who Has These Weapons?   19
    Chapter 3:   Threats and Responses   63

PART TWO: CHEMICAL AGENTS   85

    Chapter 4:   Basic Concepts   87
    Chapter 5:   Chemical Warfare: A Brief History   127
    Chapter 6:   Control and Disarmament   169

PART THREE: BIOLOGICAL AGENTS   191

    Chapter 7:   Basic Concepts   193
    Chapter 8:   Biological Warfare: A Brief History   219
    Chapter 9:   Control and Disarmament   237
    Chapter 10: Vaccination and Biological Warfare   249

Contents

PREFACE   XVII

ACKNOWLEDGMENTS   XIX

INTRODUCTION   XXI

How This Book Is Organized   XXII

PART ONE: GAS, BUGS, AND COMMON SENSE   I

Chapter 1: The Fog of War   3

GAS AND BUGS   5

CBW, Briefly Defined   6

THE UTILITY OF CBW AGENTS   7

Making the Enemy "Suit Up"   7

Leaving a Large Footprint   8

Implying a Threat   9

ACQUISITION   9

Obtaining CW Precursors   10

The Development of BW Programs   10

Dual-Use Technologies in BW   12

Production   12

CW Agent Production   12

BW Agent Production   13

WEAPONIZATION   14

CW Weapon Design   14

Weaponizing BW Agents   16

DELIVERY 16

Dispersal 16

Atmospheric Conditions 17

Wind Speed and Turbulence 18

Heat 18

Chapter 2: Who Has These Weapons? 19

THE SUPERPOWER AND FORMER SUPERPOWER 21

The United States 21

Chemical Weapons 21

CW in World War II 23

CW During the Cold War 24

Current Status of US Chemical Weapons 28

Biological Weapons 30

The Post-World War II Era and the Korean War 30

Russia and the Former Soviet Union 31

Chemical Weapons 32

The Russian Federation in the 1990s 33

Biological Weapons 34

THE MIDDLE EAST 36

Iraq 36

Chemical Weapons 36

The Gulf War (1990–1991) 36

Iraqi VX 39

Biological Weapons 40

Iran 42

Chemical Weapons 42

Biological Weapons 43

Syria 43

Chemical Weapons 43

Biological Weapons 45

Egypt 46

Chemical Weapons 46

Biological Weapons 47

Libya 48

Chemical Weapons 48

Operation at Tarhunah 48

Biological Weapons 49

Israel 49

Chemical Weapons 49

Biological Weapons	49
EAST ASIA	50
North Korea	50
Chemical Weapons	50
Suspected CW Arsenal	51
Biological Weapons	52
South Korea	53
China (PRC)	54
Chemical Weapons	54
Biological Weapons	56
Taiwan	56
OTHER PLAYERS	57
South Africa	57
Chemical Weapons	57
Biological Weapons	58
Cuba	58
SUB-STATE ACTORS	58
Chapter 3: Threats and Responses	63
A NEW KIND OF WARFARE	63
The First World Trade Center Bombing	64
Aum Shinrikyo	64
Taking the Toll	66
MEDICAL THREATS AND RESPONSES	67
Anthrax	67
The Disease	68
Anthrax as a Weapon	68
Treatment and Vaccination	69
Smallpox	70
The Disease	70
Smallpox as a Weapon	70
Vaccination	71
CIVIL DEFENSE THREATS AND RESPONSES	72
The Chemical Industry	72
Bhopal	73
Densely Populated Spaces	74
Water Supplies	74
Crop Dusters	79

Food Security   80

DETECTION   80

PART TWO: CHEMICAL AGENTS   85

Chapter 4: Basic Concepts   87

WHAT CHEMICAL WEAPONS ARE NOT   88

WHAT CHEMICAL WEAPONS ARE   88
  Properties   89
  Delivery Systems   90

BASIC CLASSES OF CW AGENTS   92
  Choking Gases (Lung Irritants)   92
    Chlorine   93
    Phosgene   95
    Diphosgene   96
    Chloropicrin (or Chlorpicrin)   96
    Ethyldichlorarsine   97
    Perfluoroisobutylene (PFIB)   97
  Blister Agents (Vesicants)   98
    Mustard (Sulfur)   98
    Nitrogen Mustard   102
    Lewisite   102
    Phosgene Oxime ("Nettle Gas")   103
    Phenyldichlorarsine (PD)   104
  Blood Agents   105
    Hydrogen Cyanide: Instrument of the Shoah   105
    Cyanogen Chloride   108
    Arsine (Arseniuretted Hydrogen)   108
    Carbon Monoxide   108
    Hydrogen Sulfide ("Sour Gas")   109
  Nerve Agents (Toxic Organophosphates)   109
    Nerve Agent Proliferation   111
    Dynamics of Nerve Agent Poisoning   111
  Incapacitants: Psychoactive Chemicals in War   112
    Belladonna, or Glycolate Alkaloids   113
    3-Quinuclidinyl Benzilate (BZ)   113
    Ergot and Lysergic Acid Diethylamide (LSD)   114

Mescaline and Its Derivatives (Phenyl Ethylamines)   116
Methaqualone   116
Harassing or Riot-Control Agents (RCAs)   116
Lacrimators (Eye Irritants)   118
Sternutators   121
Vomiting Agents   122
Banned RCAs   122
Herbicides   123
Obscurant Smokes   124
Napalm   124
Malodorous Concoctions and Masking Agents   125

Chapter 5: Chemical Warfare: A Brief History   127

FROM GREEK FIRE TO THE *FLAMMENWERFER* . . .   128
The Nineteenth Century   131
The Dawn of Organic Chemistry   133
From the *Flammenwerfer* to the Livens Projector:
The Buildup to War   136
The Livens Projector   138
Chemistry That Changed the World   140

WORLD WAR I   142
The Chlorine Attack at Ypres   143
Mustard Enters the War   144
Weapons Used and Abandoned   146

THE AFTERMATH: PERSPECTIVES ON CHEMICAL WARFARE   148
Tukhachevsky and the War Against the Peasants   150
The Wushe (Paran) Incident):
The First Use of Chemical Weapons in Asia?   152
Ethiopia: 1935–1936   152

WORLD WAR II   153
The Sino-Japanese War   154
United States and CW Policy   155
Churchill and Chemical Weapons   155
The Bari Incident   159

FROM KOREA TO THE GULF WAR   160
Allegations of Chemical Warfare in Korea   160
Yemen: 1963–1967   161
Southeast Asia: 1965–1975   161

Iran-Iraq War: 1980–1988   162
  Agents Used Against the Kurds by Iraq   164
  Iranian Chemical Weapons Development   164
Lessons from the Gulf War   166
  Khamisiyah and Sarin Release in the Gulf War   167

Chapter 6: Control and Disarmament   169

HISTORICAL PRELUDES   169
Early Twentieth Century Negotiations   169
  The Hague Conferences   170
  The Washington Arms Conference   172
  The 1925 Geneva Protocol   173

THE CHEMICAL WEAPONS CONVENTION (CWC)   175
Controlling Agents and Precursors   177
Scheduling Agents and Precursors   177
  Schedule 1 Agents and Precursors   177
  Schedule 2 Agents and Precursors   178
  Schedule 3 Agents and Precursors   179
Declarations and the CWC   179
  Export Controls   180
  The Australia Group   180
Verification of Compliance   181

MONITORING AGENTS AND THEIR PRECURSORS   182
  Monitoring of CW Agents (Schedule 1)   182
  Monitoring of CW Precursors (Schedule 2)   183
  Monitoring of Commercial Chemicals (Schedule 3)   183
Proliferation Signatures   183
Challenge Inspections   184
Managed Access   185

DESTRUCTION OF CHEMICAL WEAPONS   186
Not in My Backyard   186
The Destruction and Conversion of Facilities   188
Brain Drain in the Former Soviet Union   188

PART THREE: BIOLOGICAL AGENTS 191

Chapter 7: Basic Concepts   193

BIOLOGICAL WARFARE AGENTS   196

THE NATURE OF INFECTIOUS DISEASE   198
The Germ Theory of Disease   198
The Advent of Modern Microbiology   199
The Airborne Origin of Infectious Disease   200

DIFFERENTIATING AMONG PATHOGENS   200
Bacteria   200
The Rickettsiae   200
Viruses   201
Bioaerosols   202

BIOLOGICAL WARFARE AGENTS   204
Bacteria   205
Anthrax   205
Plague   206
Tularemia   207
Glanders   207
Q-Fever   208
Cholera   208
Viruses   209
Smallpox   209
Hemorrhagic Fever Viruses   210
Venezuelan Equine Encephalitis (VEE)   211
Foot-and-Mouth Disease   212
Biological Toxins   213
Mycotoxins   213
Fungi (Molds)   214
Botulinum Toxin   214
Staphylococcal Enterotoxin Type B (SEB)   215
Ricin and Saxitoxin   216
Trichothecene Mycotoxins (T2)   216

BIOREGULATORS   217

PROTOZOA   217

Chapter 8: Biological Warfare: A Brief History   219

BIOLOGICAL WEAPONS IN ANCIENT TIMES   219
Biological Warfare in the New World   220

BIOLOGICAL WARFARE IN MODERN TIMES   222
World War I   222
The Geneva Protocol of 1925   223
Japanese BW, 1932–1945   224

THE UNITED STATES BW PROGRAM   226
Development Phase: 1939–1950   226
The Post-World War II Era and the Korean War   229
US Testing Activities, 1951–1969   231

SOVIET BIOLOGICAL WEAPONS: 1919–1989   233
The Soviet Renaissance (1973–1989)   233
The Biopreparat Complex   235
BW in Russia Today   235

Chapter 9: Control and Disarmament   237

HISTORY   237
Summary of the BTWC   238
The Review Conferences   240
    1980   240
    1986   243
    1991   243
    1997   244

THE BTWC TODAY   245

THE FUTURE OF THE BTWC   248

Chapter 10: Vaccination and Biological Warfare 249

DISEASE AS DETERRENCE 249

Smallpox, Variolation, and the First Vaccines 250

Jenner's Vaccinia 253

Typhus and DDT 255

MODERN MILITARY VACCINATIONS 256

Typhoid and the Boer War (1899–1902) 257

Yellow Fever as a BW Threat 257

Japanese B Encephalitis and the War in the Pacific 258

Botulinum and D-Day 259

Plague and the Vietnam War (1965–1975) 259

Botulinum Toxoid and the Gulf War (1991) 260

Vaccinating for Anthrax in the Twenty-First Century 261

Historical Development 261

The Current Controversy 263

NOTES 267

SELECT BIBLIOGRAPHY 293

INDEX 295

Preface

As these lines are being written, firemen, police officers, and a host of other rescue workers are still trying to save victims of the terrorist attacks on the World Trade Center and the Pentagon. Even without precise counts of the dead and wounded in Pennsylvania, Virginia, and New York, we can already conclude that this attack was a signal event of mass destruction, with the outcome certain to make it the deadliest single terrorist act ever committed against United States citizens. The number who died in the attacks and the 100-minute aftermath will almost certainly exceed the number of American armed forces killed at Pearl Harbor, or on D-Day. But these victims of terror were, by and large, civilians, people going about their business. And the instruments of their death, the murder weapons, were not, until yesterday, considered weapons at all.

It is always hard, but especially under these circumstances, to talk about weaponry, the tools of war and terror. What can these new weapons be compared to? How can they be described? The US Department of Defense categorizes the most deadly kinds of armaments as "weapons of mass destruction," or WMDs, and defines them as "capable of a high order of destruction . . . of being used in such a manner as to destroy large numbers of people."* This does not tell us much. How does one define "a high order of destruction" and "large numbers of people"? Timothy McVeigh, who was responsible for bombing the Alfred P. Murrah Federal Building in Oklahoma City in 1995, was indicted on US federal charges for using a WMD, a truckload of improvised explosives. In the blast, 168 people were killed—again overwhelmingly civilians, men, women, and children going about their business.

Was McVeigh charged with using a WMD because 168 is a large enough number? Is this or a number near it the dividing line between an ordinary act of savagery and one in which we call the killers' implements weapons of mass

---

* Department of Defense, *Department of Defense Dictionary of Military and Associated Terms* (Washington, DC: Joint Chiefs of Staff/US Government Printing Office, Joint Pub 1-02: March 23, 1994): p. 412.

destruction? Chemical and biological armaments have the potential to kill huge numbers of people, many times the number killed by McVeigh's bomb. And certainly it seems to make sense that the US Department of Defense categorizes them as WMDs. But is it not true that almost any weapon, even the machetes used in Rwanda in 1994, can be used to perpetrate horrors on an unspeakable scale?

The term "weapons of mass destruction"—probably coined in 1956 by the Soviet Red Army Marshal Georgi Konstantinovich Zhukov (known as "the hero of Stalingrad")—has, like any defining or categorizing word, its shortcomings. It explains some things, but goes only so far. The arms expert Ken Alibek, whom we shall meet later in this book, suggests that a better name for biological armaments might be "mass casualty weapons," since their object is to inflict human injury but not to destroy buildings or property. Distinctions like these are grim—but they are also useful. They help us refine and sharpen our sense of things. They help us face up to and describe in words what otherwise may be overwhelming, confusing, frightening. And of course trying to face up to facts and describe events—no matter how horrible they may be—is the first step toward understanding.

My wish is that readers will take up this book in that spirit. Studying weaponry and warfare and disarmament isn't just a challenging and stimulating intellectual discipline for its own sake. The stakes are much too high for that. Its aim instead is to help us understand a long-standing aspect of human behavior, a force in human history, that seems capable of devising new tools of destruction that we may have to face at times and in places where we least expect them—in a pair of towers above a great harbor, in offices at the heart of our vast and powerful military establishment, and in a quiet country field in southwestern Pennsylvania.

Eric Croddy
Monterey, California
October 23, 2001

Acknowledgments

The Monterey Institute, Center for Nonproliferation Studies, would like to thank the Ploughshares Fund for their generous support of this project. Any views expressed herein are solely those of the author, and not necessarily those of the Ploughshares Fund or the Monterey Institute.

I would personally like to thank the following people, all of whom were instrumental in making this book happen. First, I want to thank Jonathan B. Tucker, the director of the Chemical and Biological Weapons Nonproliferation Program at the Monterey Institute, who initiated this project and Clarisa-Perez Armendariz and John Hart who made strategic contributions towards the development of the text. I also wish to thank other colleagues at the Monterey Institute were always there for me when I needed assistance or scholarly advice: Jason Pate, Diana McCauley, Gavin Cameron, Kathleen Vogel, Gary Ackerman, Kimberly McCloud, Raymond Zilinskas, Amy Sands, Tim McCarthy, Mari Sudo, Dan Pinkston, Amin Tarzi, Michael Barletta, Gaurav Kampani, Sonia Ben Ouagrham, Fred Wehling, Yuan Jing-Dong, Evan Medeiros, Lisa Burns, Phillip Saunders and Clay Moltz. Dr. Anthony T. Tu has been most helpful, and I am privileged to have his friendship as well as expertise. Martin Hugh-Jones has been most patient with my inane questions from time to time. My good friends and scholars at the National Chengchi University in Taiwan, Yuan I and Arthur S. Ding have taught me a great deal about East Asian proliferation issues, and Jean-Pascal Zanders has been a wealth of wisdom and a dear friend. Thanks also go to my colleagues and Monterey Institute of International Studies graduates Ken Palnau, Anisha Lal, Anjali Bhattacharjee, Matt Osbourne, Tim Ballard, Faith Stackhouse, Garvey McIntosh, and Jason Evans for their invaluable support in developing the book. My dear friend Al J. Venter has made important contributions to my thinking about a number of issues discussed here. Special thanks go to the excellent editorial staff at Copernicus Books for really making it all happen and to Michael Hennelly for his invaluable advice. Our appreciation goes to Pernacca Sudhakaran, Ewen Buchanan, and the excellent staff at the UN photo laboratory for their dedication and generous use of photographs. There are many others who would prefer to remain anonymous, but I want to thank them as well.

A big thank you to my family for their love and support.

Any errors, committed or otherwise, are solely my responsibility.

Introduction

Why study chemical and biological warfare (CBW)? At the very lowest level, the topic lends itself to morbid curiosity. The scale on which "bugs and gas" can be used to kill people, and the way in which they cause death, can make for gruesome reading. Then there is the matter that these weapons are considered, rightly or wrongly, to be abominable, and those who wish to confirm that opinion will find in studying CBW plenty to abhor. Readers in these two categories are likely to be disappointed by what they will find in this book.

Fear is another motive for study. One can hardly read the paper or listen to the news today and not, sooner rather than later, hear reports about the belligerent nations, repressive regimes, and terrorist organizations that have access to, or are working on the development of, these weapons. The mere existence of CBW armaments, we are told, poses a significant threat to the stability of international order. Even if one believes that the nuclear stand-off between superpowers—the Balance of Terror that characterized the Cold War—is a thing of the past, we now have a whole new cast of characters to worry about. They are less well understood than our old adversary the Soviet Union, and less predictable. They operate as states (or sometimes "rogue states"), but also in the shadows, in league with networks of terrorists, global criminal enterprises, and splinter groups representing every conceivable type of fanaticism. And they will, it is almost certain, push us into a whole new kind of decades-long war. For readers arriving with this point of view, I hope this book will serve as a kind of corrective.

It is not my belief that CBW armaments are benign, or that states and sub-state organizations are not wishing for or even planning chemical or biological attacks against the United States and the rest of the industrialized world. I am not someone who places great faith in the good will and sober judgment of, say, Saddam Hussein. In fact, if I were a betting man, I would put my money on the likelihood that we *will* see chemical or biological weapons attacks in the not-too-distant future. But where this book perhaps differs from some more popular discussions of the topic is in its argument, in its underlying theme, that biological and especially chemical attacks of any magnitude are extremely difficult to plan, develop, execute, and fund. Certainly it is true that a fanatical cult could release nerve agent on a crowded subway car, as happened in Tokyo in March 1995. And the ultimate splinter group, a single deranged individual, may be perfectly capable of killing, injuring, or incapacitating large numbers of individuals in any number of ways chemical or biological. If you add to these all the belligerent major powers, rogue states, and oppressive regimes world-wide (and factor in their client terrorist organizations as well), you can imagine no end of mischief—gas attacks, reservoir poisonings, anthrax outbreaks, and so forth. But what we have to do is dwell less on nightmare scenarios and try to learn—as calmly and clearly as possible—what CBW agents are, how they work, who has used them in the past, and what is being done to limit their proliferation. Fear may be a good motivator, but it is not, as far as I can tell, an aid to understanding.

How This Book Is Organized

This book is divided into three major parts. In Part I, "Gas, Bugs, and Common Sense," there is a brief introduction to and definition of CBW (Chapter 1), including descriptions of why and how nation-states and "sub-states" (for example, terrorist organizations) develop chemical and biological weapons. Chapter 2 then lists, in a fairly straightforward manner, the nations that have CBW capabilities, along with brief descriptions of the particular agents they possess. In Chapter 3, we take a look at some of the threats we're likely and unlikely to face.

Part II is focused on chemical weapons. In Chapter 4, there are rather extensive descriptions and discussions of more than fifty of the best-known CW agents. Chapter 5 is a history of chemical warfare from ancient times to the present. And Chapter 6 discusses in detail the workings of the 1992 Chemical Weapons Convention (CWC), by all accounts one of the most effective international treaties written. (But not, as the chapter makes clear, without its limitations.) Included in the chapter is a lengthy discussion of the extremely difficult matter of verification, and the highs and lows of the international community's relationship with Iraq, an unwilling signer of the accord.

Part III, which more or less mirrors Part II, focuses on biological agents and weapons, with Chapter 7 describing more than forty biological agents in detail. Chapter 8 focuses on BW armaments in history, again covering a broad span. Chapter 9 covers the Biological Weapons and Toxins Convention of 1972 (BWTC), a work of the best intentions but not much good effect. (The success of the CWC and the comparative ineffectiveness of the BWTC are discussed in some detail.) Finally, a whole chapter (Chapter 10) is devoted to the issue of vaccinations and biological warfare.

PART ONE

Gas, Bugs, and Common Sense

CHAPTER 1

The Fog of War

Gas! GAS! Quick, boys!—
    An ecstasy of fumbling
Fitting the clumsy helmets just in time,
But someone still was
    yelling out and stumbling
And flound'ring like a man in fire or lime.—
Dim through the misty panes
    and thick green light,
As under a green sea, I saw him drowning.
In all my dreams before my helpless sight
He plunges at me,
    guttering, choking, drowning.

—*from "Dulce et Decorum Est,"*
by Wilfred Owen, 1918[1]

War has always been a nasty business, but to most of us, war waged with gas—with poison vapors, with clouds of toxins—holds a special horror. When we try to imagine (and most of us have only imagined) attacks with such weapons, we don't picture heroic struggles, or valiant combat, or noble death. Even though accounts of such battles demonstrate that there was no shortage of heroism, valor, and nobility, there remains at least in the popular imagination a picture of soldiers engulfed in a different and especially terrifying fog of war—blinded, disoriented, groping in the obscurity, clutching their chests and gasping for breath. There is in the experience, as Wilfred Owen helps us imagine, something of the horror of drowning—only worse. Many of these weapons, we know, cause asphyxiation. Their victims choke to death.

Then there is the matter of silence. A chemical or biological agent can, to be sure, be targeted and delivered in an artillery shell or explosive bomb, but it may just as well be quietly released from a steel canister, as was the case in the

Various masks, some of dubious utility, used throughout World War I. Top row, from left to right: US Navy Mark I mask, US Navy Mark II mask, US CE mask, US RFK mask, US AT mask. Middle row, from left to right: British Black Veil mask, British PH helmet, British BR mask, French M2 mask, French artillery mask, French ARS mask. Bottom row, from left to right: German mask, Russian mask, Italian mask, British Motor Corps mask, US Rear Area mask, US Connel mask. (Courtesy of Soldier Biological and Chemical Command, Historical Research and Response Team, Aberdeen Proving Ground, MD.)

earliest gas attacks in World War I. There must be an eerie quality to being under attack when there is no muzzle flash, raised bayonet, loud explosion, or charging infantrymen. To the ordinary horrors of battle is added the special terror of an attack that is not just silent (or nearly so) but invisible (or nearly so). A gas attack can be underway long before we know it, and an attack with germs may be days in the past before symptoms begin to appear. The enemy, the human agent responsible, will very likely not be present, but nameless, faceless, and long gone from the scene.

Chemical and biological warfare, or CBW, seems somehow "modern" to us, but modern in a thoroughly bad way. Many historians have pointed out that the twentieth century began not in January 1900, but August 1914. The modern era, so goes the thinking, truly started with the beginning of World War I. It was the war that depersonalized killing, mechanized killing, industrialized killing on a massive scale. The total dead, the eminent historian of warfare John Keegan tells us, was something on the order of 20 million people.[2] Above all else, World War I, the Great War, lives on in memory not as the war to end all wars, but the war in which soldiers lay gassed in trenches, and the war that gave us a foretaste of the even more murderous conflicts the twentieth century had in store.[3]

GAS AND BUGS

As it turns out, in the history of twentieth-century warfare, chemical and particularly biological weapons—known colloquially as "gas and bugs"—have had a very small impact on the toll of dead and injured. Even more so than nuclear weapons, they have seen only very limited use, and the number of fatalities that can be directly attributed to them pales in comparison to the deaths and dismemberments effected with artillery, rifles, bombs, torpedoes, and the rest of the conventional arsenal. Also like nuclear weapons, chemical and biological weapons terrify us in part because they have *not* been widely used. They exist as *potential* threats. Because they have not seen frequent and repeated deployment, their effects in real battle or in widespread terrorist attacks can only be guessed at. Of course our defense and intelligence agencies have researched and tested a good many chemical agents (and somewhat fewer biological agents). They have drawn up models and tested hypotheses on the laboratory bench and in the field. They have closely examined the evidence from the few actual attacks that have taken place in the world outside. But the fact remains, part of the fear we have of chemical and biological agents is that they are simply not tried and true. Our lack of experience with them actually adds to our uneasiness.

CBW, Briefly Defined

If we want to understand these weapons and their intended uses, we need to do more than fear them, and the place to begin is with definitions. CBW, or chemical and biological warfare, is waged with chemical and biological *agents* that have been placed in weapons. (In the language of the military community, the agents have been *weaponized*.) Chemical agents are made up of what are called *precursors*—simply speaking, their ingredients. Biological agents are *pathogens*, disease-causing organsims or substances.

Chemical warfare (CW) weapons employ poisons that kill, injure, or incapacitate. The word "poison" does not necessarily mean a substance that will cause death, but it does imply that a small quantity is sufficient to have a harmful effect—on a human, animal, or plant. In common usage, chemical weapons are often thought of as "gas," and to this day, soldiers in even the most modern armies are instructed to raise the alarm "Gas!" to indicate that an attack with chemical weapons is underway. But almost no modern chemical weapons are actually gaseous at normal temperatures and pressures. For example, mustard and sarin are two highly toxic chemical agents that exist principally in liquid form. The most effective way to achieve high concentrations of these agents in an attack is in the form of aerosols—very fine, suspended particles that remain airborne for a signifacant period of time. In the parlance of CBW, the designers of chemical weapons have to carefully plan *delivery* strategies, determining the most effective form in which to get the agent from the weapon to the target.

Biological warfare (BW) weapons make use of an agent that contains living organisms (such as the bacterium that causes anthrax), viruses (like the one that causes smallpox), or toxins that have been harvested from microbes or extracted from plants. Sometimes the distinction between chemical and biological agents is not clear. Ricin, for example, is a toxin extracted from the castor bean plant but could easily be considered a chemical weapon, and in a broad sense any toxin from any organism is still essentially a chemical. Still, in general usage and for the purposes of this book, we will use the label "biological" to refer to agents that use living organisms and their toxins in order to inflict injury and death. Like chemical agents, biological agents have to be weaponized and designed into efficient delivery systems.

Later, we'll take a much closer look at chemical agents (in Chapter 4) and biological agents (in Chapter 7), and refine and qualify these definitions. For now, though, it is enough to know that given sufficiently harmful agents which are effectively weaponized and designed for efficient delivery, an army or a terrorist organization or even an individual can use chemical and biological agents to do great harm. The object of the attack can be almost any living thing—an army or a civilian population, livestock or work animals, crops or jungle cover—but the goal is the same: to make sure the agent is inhaled, or ingested, or makes surface contact in quantities that will incapacitate, injure, or kill.

THE UTILITY OF CBW AGENTS

The number of nations that have developed or tried to develop CBW capabilities has grown alarmingly in the last 25 years. So, too, has the number of "substate" organizations, including factions in civil wars, terrorist organizations, and religious cults—and even, we would hazard a guess, deranged individuals. At the risk of posing a question with an obvious answer, we should ask why have they worked so hard to secure these weapons? Clearly, the weapons are useful in killing and injuring and incapacitating one's enemies. No one can dispute that. And then there is the argument, also hard to counter, that if our enemies have these weapons, we must have them too. Almost no one, it would appear, wants to be left unable to respond in kind. But what is it about these armaments in particular that makes them different from, and in some ways more effective than, the ordinary implements of chaos, mayhem, and death?

During the trench warfare that dominated World War I, the immediate incentive for one of the armies to employ chlorine gas was to break through the sturdy trench fortifications that had withstood attacks with conventional weapons. Heavier-than-air chemicals offered a means to break the lines of even the best dug-in defenders. The goal became to injure or kill large numbers of entrenched forces, then overrun their positions.

Today, mobile warfare, with air support and more powerful and accurate conventional weapons, has made this motive for using gas all but obsolete. No system of trenches could withstand the attack of a modern army, and indeed no modern army would think of placing itself in fixed entrenchments. So what then is the appeal of CBW to the modern warrior?

Making the Enemy "Suit Up"

Military strategists often refer to "force multipliers," tactics or matériel that can dramatically improve one's position on the battlefield. Early on, it was widely recognized that forcing an enemy to "suit up" with chemical protective gear was a highly desirable goal. Suiting up is an expensive and, more importantly, time-consuming process. It adds to the chaos and confusion and miscommunication—the infamous "fog of war" that inevitably descends on an active battlefield—degrading the overall fighting ability of the troops, distracting commanders' attention from other urgent tasks, making even the simple shouting of a command, not to mention radio communications and computer operations, more problematical. (Entering data on a keyboard is not easy when you're gloved and hooded.) Moreover, the mere act of donning full chemical and biological warfare protective gear can also cause·a small but significant percentage of personnel to panic, and in the ensuing chaos and confusion can lead to the abandonment of positions and weapons, a loss of focus, and failures in communication and command structures.

While CBW weapons are primarily chosen because of their extreme toxicity (or infectivity), in the eyes of the strategist, lethality is not necessarily the primary or even a desired goal. Creating mass enemy casualties—that is, injured personnel but not necessarily fatalities—forces the opponent to spend precious time and resources to care for the wounded. Similarly, in most instances of terrorism, it is the violent nature of the act itself—sometimes even more than the casualties and fatalities involved—that strikes fear into the heart and mind, that in other words *terrorizes.*

So there are non-lethal benefits, if that word can be used, to employing chemical and biological weapons against your enemy: You will confuse him, and slow him down, and disorient him, and make it harder for him to communicate. It is in the very nature of these weapons, as we have said, to increase uncertainty, and to spread fear.

Leaving a Large Footprint

CBW weapons have another advantage. They come at a relatively low cost, considering the size of their "footprint." When compared to the cost of conventional modern weapons, and especially when compared to the investment required for nuclear war, chemical weapons are cheap, and biological weapons are even cheaper.

To be sure, developing and deploying a significant amount (say, a thousand tons or more) of a chemical agent or hundreds of tons of a biological agent is an extremely complicated and expensive exercise. If the agents are to be used in battle on a large scale, one has to invest in the mass-production of sophisticated munitions (fuses, casings, etc.). There are great hazards involved in filling bomb or shell casings with deadly poisons, germs, or toxins. And even if the goal is to use chemical or biological weapons on a smaller scale, there is the matter of weaponizing the agents, and training the personnel who will handle them. Even if a state or a terrorist organization is willing to dispense with protective clothing, decontamination equipment, and detection devices to shield its own personnel, it at the very least has to train them and outfit them with the tools they need to deliver the agents effectively. Once released, particularly by inexperienced or poorly trained personnel, these agents can, with a shift in the wind or a quick change in the temperature, actually turn back on the attackers.

Implying a Threat

It could be said that a nation-state, or perhaps even a terrorist organization or religious cult, when armed with chemical and biological weapons, will neither be trifled with nor ignored on the world stage. The real or perceived benefit is that, once accepted as a player in the CBW business in a serious way, almost any organization has strategic clout. And given the nature of these weapons, and under the right circumstances, an organization of almost any size, or a nation in almost any state of economic and military and diplomatic collapse, can mount a chemical or biological attack. The statement that chemical weapons are a kind of "poor man's atomic bomb"[4] was made by Iranian President Ali Akbar Hashemi Rafsanjani in 1988, when Iran was near the end of a grueling and murderous war with Iraq, and was reeling from the losses it had suffered. It was perhaps a boast about Iran's capabilities, but it was also a threat of reprisal, and no matter how beleaguered Iran was at the time, it could not be ignored. The threat of CBW weapons can be used as well for deterrence, as in the case of Syria's deployment during the 1990s of chemical weapons to deter a possible preemptive Israeli nuclear strike (and perhaps protect Syria's own atomic bomb program). At the same time, despite Iraq's highly publicized threat to retaliate with chemical weapons, Baghdad's CW capability did not deter the Allied Coalition from its 1991 attack and defeat of Iraq in the Gulf War.

But with the evidence of threats already expressed by states and rogue regimes, by worldwide terror organizations and tiny splintered cults, it is clear that chemical and biological agents as mere threats are powerful tools in the hands of those who feel themselves to be dispossessed, defeated, overwhelmed, or outgunned.

ACQUISITION

To make good on threats, or at least to back up boasts of deadly CBW capabilities, a state or sub-state organization needs to acquire chemical and biological weapons either by purchasing them or by developing the weapons themselves. In most cases, the process of acquisition involves both, combining the home-grown, so to speak, with the store-bought.

Some CBW "proliferators," the states and organizations currently building up chemical and biological arsenals, obtain their weapons, expertise, and technology from abroad. Others rely on indigenous materials and talent. In either case, though, the costs can be high.

Obtaining CW Precursors

Chemical warfare precursors can either be bought from outside suppliers or synthesized from raw materials to which the proliferator has access. The fact that many chemicals used in civilian industry can also be employed to make CW agents has contributed to the worldwide proliferation of chemical weapons. These chemicals are called "dual-use" compounds since they have legitimate commercial uses as well as applications in the production of armaments. (See Table 1-1.)

Table 1-1. Dual-Use CW Compounds[5]

Compound	Commercial Use	CW Use
Thiodiglycol	Plastics, textile dyes, ink	Mustard
Phosphorus trichloride	Plasticizers, insecticides	G-series nerve agents
Sodium cyanide	Dyes, pigments, metal hardening	GA, AC, CK
Methylphosphonic difluoride	Organic chemical synthesis	VX, GB, GD
Phosphorus pentasulfide	Lubricants, pesticides (e.g., Amiton)	VX

While unrestricted trade in these compounds was common in the past, voluntary efforts and international agreements like the Chemical Weapons Convention of 1993 (see Chapter 6) have attempted to make export controls consistent globally and to restrict the availability of many such precursors. But needless to say, not all chemical dealers honor these arrangements, and even if they did, proliferation would not stop altogether. As stated before, proliferators may manufacture precursor chemicals from simpler compounds whose export is not controlled or from compounds that are available from domestic sources. This is called "back-integration." Iraq, for example, applied back-integration to mustard gas production in the early 1980s. Unable to make thiodiglycol, an immediate precursor of sulfur mustard, it ordered the precursor from foreign sources. Then, when an embargo threatened the supply from Western countries, Iraq developed an indigenous production capability based on reacting ethylene oxide with hydrogen sulfide.

The Development of BW Programs

While the minimum level of infrastructure, equipment, and technology required to develop biological warfare capability are relatively inexpensive and accessible, there are significant technical barriers to acquiring a sophisticated and reliable BW program.

Specialized facilities of some sort are required for the production of BW agents, if only to maintain safety for workers. Research involving very infectious organisms often takes place in a containment facility with a high biosafety level (BL), the highest being Level 4. BL-4 laboratories would be the best equipped to handle the most dangerous pathogens currently known. It is

by no means necessary, however, to possess sophisticated containment structures to develop a biological weapons arsenal. The production of anthrax cakes
during World War II in England, for example, entailed no protective procedures more sophisticated than rubber gloves and aprons with detachable
sleeves, and vigorous washing.[6]

To begin a BW program, a proliferator must acquire a seed culture of the
agent it intends to produce. Some disease-causing pathogens and toxins occur
naturally in the environment. For example, the bacterium that causes anthrax,
Bacillus anthracis, can be found in soil worldwide, and manifests itself primarily
as a disease of farm animals (cattle, horses, and sheep, etc.). When animals
become infected, anthrax bacteria show up in their hides and carcasses, making farm workers, veterinarians, and meat cutters particularly liable to infection. (People can also become infected by anthrax in airborne form; this will
be discussed in detail in Chapter 7.) Some small quantity of anthrax bacilli
could, conceivably, be isolated from a natural starting place such as a farm and
then be used as a seed stock, but it is unlikely that a proliferator would take
this route. Isolating the microbe, then processing it and finally culturing it to
produce sufficiently large quantities all involve extremely difficult processes,
and the proliferator would want to make sure that the particular strain of
anthrax he had was the hardiest and most virulent. A better way of securing
the right stock would be to steal it from a research institute, public-health
facility, hospital, or university laboratory, all of which may keep the bacterium
in a variety of well-defined strains for purposes of diagnosis, testing, and
research.

Today, there is a coordinated effort among many nations to guard the stocks
of these germs more closely, in the hope that fewer proliferators will have easy
access to seed stocks. (The Biological and Toxin Weapons Convention of
1972, discussed in Chapter 9, was an early attempt to stem the tide of BW
proliferation.) The problem is that rules about access are not consistently
applied by all countries and companies. Moreover, efforts at tighter control
may have come too late, since as recently as just a few years ago, samples that
could be used as seed stocks were available on the open market from reputable
suppliers. In 1986 and 1988, for example, Iraq purchased pathogens for
anthrax, botulism, and gangrene, among others, from the American Type
Culture Collection, a world-renowned not-for-profit company in the business
of providing germ specimens for researchers and hospitals.[7] (See Table 1-2 for
a list of the pathogens Iraq purchased.) And in late 1998, British journalists
posing as Moroccan scientists were able to find at least one firm in the Czech
Republic willing to sell them samples of a botulism toxin for 50 German
marks, or about $25, each.[8]

**Table 1-2. Iraqi Biological Specimens from
the American Type Culture Collection**[9]

Microbe	1986	1988
Bacillus anthracis (anthrax)	†	†
Bacillus megaterium	†	
Bacillus subtilis	†	
Bacillus cereus		†
Brucella abortus	†	†
Brucella melitensis	†	
Clostridium botulinum (botulinum toxin)	†	†
Clostridium perfringens (gas gangrene)	†	†
Clostridium tetani	†	
Francisella tularensis (tularemia)	†	

Dual-Use Technologies in BW

The problem of dual-use technologies, other than dual-use precursors, is what challenges those investigating the proliferation of biological weapons. A proliferator can mass-produce BW agents in, for example, a large fermenter that one would find in a bona-fide brewery. Or a test of how well an agent could be aerosolized might be conducted using agricultural fertilizer sprayers at an actual working farm. Much of the equipment, knowledge, technology, and infrastructure required in agriculture and medicine can be put to peaceful uses as well as non-peaceful ones. Moreover, newly advanced techniques for sterilizing contaminated equipment now help BW proliferators to do a better job converting equipment back to its legitimate purpose. The dual-use problem means that even though BW proliferation is prohibited by international law, investigators may have a very hard time proving definitively that a particular facility was used to test or develop these armaments.

Production

The production of CBW agents is fraught with technical difficulties, and before mass-production of agents and weapons begins, smaller operations involved in pilot production, sometimes taking place in nothing more elaborate or extensive than a laboratory setting, are necessary to test materials and techniques.

CW Agent Production

The first step in production involves the synthesis of chemical warfare agents in small quantities at a pilot facility in order to iron out technical details. Especially when novel approaches are being taken, small batch production lines are needed to fine-tune the process. In the course of manufacturing CW agents, technicians and workers often have to learn how to deal with extremely harmful and dangerous by-products and their environmental consequences, and a pilot program is the right place to learn.

In some ways, the scaling up to mass-produce CW agents is no different from scaling up an operation that makes legitimate commercial compounds. Both involve the use of standard chemical process equipment, including reactor vessels, in which synthesis is actually carried out; heat exchangers to control temperature; and various pumps, pipes, valves, and joints. Double-seal pumps, special air handling systems, filters, and other devices designed to handle exceptionally toxic compounds are also desirable. Access to all of this equipment, and maintaining it and replacing it, can be a special challenge for clandestine operation.

While standard chemical process equipment might be suitable for a *sub rosa* CW program, specialized equipment is preferred because the compounds involved are often highly corrosive. This is especially the case in the production of nerve agents such as sarin and soman. Nonetheless, Iraq seems to have utilized non-specialized steel reaction vessels for its VX manufacturing process, an instructive example of how proliferators may disregard or stretch the definition of safety precautions in order to hide the real purpose of certain facilities. Scott Ritter, a former US Marines intelligence officer and inspector with UNSCOM, a UN-based organization verifying compliance with the UN resolutions related to the Chemical Weapons Convention, described how difficult it is to keep pace with what goes on in the real world:

> ...The method used by Iraq to manufacture stabilised VX [a
> chemical nerve agent] was done in stainless steel reactor vessels.
> But UNSCOM currently monitors only glass-lined reactor vessels
> ... largely because nobody thought stainless steel vessels were
> useful and because monitoring stainless steel vessels was too diffi-
> cult. Given what we now know of Iraq's VX programme, how
> can we safely say that the Iraqi government did not produce VX
> between 1991 and 1998? For instance, the Baiji fertiliser complex
> is one about which information suggests could have been used in
> such a covert effort. ...[10]

BW Agent Production

From the proliferator's point of view, and especially in the case of smaller nations and terrorist organizations, the production of biological weapons, although fraught with all kinds of dangers, is in many respects easier, cheaper, faster—and perhaps most important, not difficult to hide.

A biological warfare program is easy to conceal because there is no need for large storage facilities. Small amounts of a BW agent seed culture can multiply into large quantities in about two weeks. In the event of a war, a country that has conducted sophisticated BW research and development may not need a long-standing stockpile of BW agents; it can produce and deploy a viable BW arsenal from a small amount of agent within a few months. Moreover, unlike chemical and nuclear weapons production plants, militarily significant amounts of BW agents can be produced in a laboratory no larger than a trailer home.

The Biological and Toxin Weapons Convention (BTWC) of 1972 prohibits the development, production, stockpiling, and transfer of biological warfare agents and the devices used to deliver them. However, the fact that BTWC permits defensive research into developing countermeasures for BW agents complicates the issues of concealment and verification. Defensive and offensive BW research involve many of the same processes, equipment, and infrastructure. Because the intent of a defensive program is to develop BW detection devices, prophylaxis, and therapy, and because the burden of proof lies heavily on proving *intent*, a proliferator could use BW defense research as a cover for offensive research.

Furthermore, because many diseases are endemic to many countries around the world, legitimate-sounding cover stories can also be developed to justify the research into plague, anthrax, tularemia, etc. Added to that is the fact that the BTWC, unlike the CWC, does not yet provide for extensive on-site investigation and verification.

WEAPONIZATION

Weaponization involves the development of technology to process and deliver the agent effectively. In the course of creating a CW weapon, it would involve everything from insuring that the agent is chemically stable over the range of environments likely to be encountered in an attack, to the actual filling of the munitions (artillery shells, rockets, simple canisters). Those agents that are gaseous at room temperature require pressurized vessels, while liquids and solids are best disseminated in aerosol form. Both chemical and biological agents must be able to withstand the trip from storage to perhaps flight at 30,000 feet, and in many cases the heat and shock from explosive dispersal. These especially apply in the case of BW agents. Also of concern are the effects of ambient conditions such as moisture and UV light at the scene of an attack.

CW Weapon Design

The weaponization of CW agents often involves the use of stabilizers to prevent the degradation of agents, often achieved by neutralizing the acids released by chemical decomposition. Some of these stabilizing compounds add more than stability to the weapons, creating higher yields, or increasing the viscosity, and thus the persistence of some agents. The nerve agent tabun, for example, the first nerve agent manufactured in mass quantities by Nazi Germany, was stored mixed with the stabilizer chlorobenzene. Other preparations (such as sarin) that are prone to being acidic can be neutralized with some kind of additive such as trimethylamine. And still other additives can enhance the physical properties of certain agents. For example, a mixture of mustard with diisopropyl fluorophosphate (DFP), itself a nerve agent, lowers the freezing point down to −36°C, making it suitable for use in weapons deployed in severe cold.[11]

The designers of CW armaments also have to consider the dispersal of the agent at the moment of impact or release, an engineering problem made difficult by the volatile nature of many substances used in chemical warfare. Most CW munitions are designed to disseminate an aerosol of microscopic droplets or particles that can be readily absorbed by the lungs, or a spray of larger droplets that can be absorbed by the skin. All of these factors need to be taken into account.

Another major issue the designer has to address is whether to build a binary or unitary weapon. So-called "unitary" munitions contain the CW agent already in its completed toxic form, while binary weapons maintain two separate chambers: one component mixes with the other to produce the CW agent before reaching the target.

Unitary munitions may take the form of a shell, rocket, bomb, or canister filled with the CW agent, and only a fuse needs to be added before firing. However, unitary munitions are dangerous to store, handle, and transport. Binary munitions, which are generally considered a significant improvement in design, reduce the likelihood of serious accidents, especially if the two components are kept apart until the last possible moment. The US 155-mm artillery munition utilized a binary system that combined the precursor chemicals difluor[12] (DF) and isopropyl alcohol to produce sarin.

A more sophisticated but also more problematic approach from an engineering standpoint was the US Bigeye bomb, or what could be more accurately called a spray tank. In this ordnance, relatively innocuous components including solid sulfur and the immediate precursor QL were set to combine and release the deadly nerve agent VX. As the bomb glided over the intended target, the aerosolized compound was sprayed in its wake.[13]

An example of a crude hybrid of unitary and binary chemical weaponry is the case of Iraqi manufacture of sarin and GF (cyclosarin) agents. By mixing a precursor (difluor) with already poured alcohols (isopropyl, cyclohexanol) in bombs or missiles, this Iraqi "quick-mix" procedure, employed just before use on the battlefield, produced an approximate 60:40 sarin/cyclosarin mixture. This method involved a reaction involving rather unstable components, and was no doubt a very risky operation.

Weaponizing BW Agents

In order to process biological agents into a viable weapon, a producer must make them capable of surviving storage and dissemination. Also, the agent must have an acceptable particle size for optimal aerosol delivery. A process known as *lyophilization*, or freeze-drying, can be used so that BW agents remain potent while being stored. Lyophilization can reduce a solution of bacteria and a sugar-like stabilizer to a small cake of dried material that can be milled into a fine powder. Anthrax spores prepared as a powder at Stepnogorsk, a BW research and production facility in the former Soviet Union, for example, were estimated to have a shelf life of more than 75 years.[14]

To create effective biological weapons, BW armament designers may also employ a technique called *microencapsulation*. This process involves coating liquid or dry BW agents and toxins with materials designed to induce slow or targeted release of an agent within the host. It can also produce agent particles of an optimal size and increase the agent's stability while retaining its potency.[15] A proliferator could benefit from microencapsulation technology in order to manufacture 1- to 5-micron particles, protecting the agent until it reaches the lower lung and the alveoli. Virulent agents that would ordinarily be vulnerable to the elements or the body's natural defenses could be optimized in order to exert maximum harm on a large number of people. However, microencapsulation is technically complex and demanding, and more study is required to understand better how microencapsulation may increase the biological weapons threat.[16]

DELIVERY

Finally there's the matter of delivery, by which we mean not merely the carrying of a bomb to the front but the method by which the CBW agent is actually dispersed.

Dispersal

There are two main delivery methods for CBW agents: Point source dispersal makes use of munitions, such as an artillery shell or missile warhead. Line source dispersal employs a sprayer system. The choice of one or the other depends on what is required for effective distribution of the chemical agent.

CW agents that easily cause death or injury by contact, such as mustard and VX, could be distributed both as large droplets (70 microns or so[17]) and depending upon the explosive charge configuration, in very small, aerosolized particles. (The thickness of a human hair is generally 75 microns or more, and with bright light contrast, airborne particles as small as 20 microns can be seen with the naked eye. See Table 1-3.) For less persistent agents such as sarin, aerosolization and vapor enhance the effectiveness of the agent, increasing field concentrations of the agent and producing particles in the 5- to 10-micron range. These tiny particles can reach deep into the lungs.

Table 1-3. Sample Particle Sizes

Substance	Diameter, Microns
Tobacco smoke	0.25
Smallpox virus	0.40
Anthrax spore (minimum)	1.00
Talc powder	10.00
Flour dust (minimum)	15.00
Flour dust (maximum)	20.00
Pollens (minimum)	15.00
Pollens (maximum)	70.00
A human hair (minimum)	75.00
A human hair (maximum)	100.00

Chemicals such as mustard (a blister agent) and sarin (a nerve agent) can form toxic vapors at room temperature. Agents that share this physical property could be employed in crude but still very deadly attacks without sophisticated dispersal devices. Other, more viscous compounds such as VX are much less prone to evaporation even at higher temperatures, and for this reason present more of a hazard as a skin contaminant.

Atmospheric Conditions

The overall effectiveness of CBW agents on the battlefield, particularly when delivered in the form of aerosols, is heavily influenced by atmospheric conditions in the target area. One and the same weapon, used on a breezy battlefield and in an air-conditioned shopping mall, will have profoundly different effects. Depending upon the physical as well as the chemical traits of a given agent, delivery systems have to be designed to take into account how the agent will react in wind, in bright sunlight, in rain, at altitude, and so forth.

In particular, patterns of heat and cold have to be considered. Inversion occurs when the ground is cooler than the air above, and as one moves higher above ground the air temperature rises. During inversion, there is very little air turbulence and low wind speeds, and subsequently CBW agents will not be dispersed and diluted very rapidly. At the same time, however, inversion tends to make aerosolized particles more prone to falling out of air. Nonetheless, inversion is considered the optimum condition for the ground release of CBW agents.

Wind Speed and Turbulence

Wind speed and direction are likewise critical factors for getting an agent from the release point to a target. Under the right wind conditions (low velocity, minimal turbulence), agents can be transported on the wind very effectively in concentrations sufficient to cause widespread casualties. A related factor is atmospheric stability. Turbulence and currents in the air are very important, establishing a "mixing layer" in which the agent is circulated. Depending upon the temperature inversion in the atmosphere, the height from the ground to the mixing layer may vary from as little as 20 meters to more than 1500 meters. Atmospheric stability and the height of the mixing layer are also very important for determining at what altitude a particular agent should be released, particularly if it is released from a plane or missile warhead.

Heat

Ambient conditions such as heat, and thus the time of day in which an agent is released, are critical elements in deployment. If released at a high altitude, agent particles may not be able to reach sufficient concentrations to inflict casualties. Humidity can also affect dispersal: some CW agent particles, absorbing moisture from the air, become too heavy and fall harmlessly on the ground, while rain can wash the air and even neutralize by hydrolysis (albeit slowly) many CW agents.

CHAPTER 2

Who Has These Weapons?

In the course of the twentieth century, dozens of nations have developed, a few have deployed, and fewer still have actually used CBW armaments. After the Armistice in 1918, the principal combatants were shattered and exhausted; they seemed to have little use for chemical weaponry. Very broadly speaking, they agreed that "gas warfare" should never again have a place on the battlefield, and even as they began to rebuild their stocks of armaments over the next ten years, they focused on conventional armaments and new tools like airplanes and submarines—but not gas. In fact, for most of the half-century after World War I, the trend in military planning (with a few notable exceptions) was to move away from chemical warfare. And the same could be said for the trend in political thinking. Justly characterized as an era of unprecedented mass killing, the twentieth century had begun with a war in which gas munitions came to symbolize all that is horrible about modern warfare, and the end of the war brought numerous international calls for disarmament, and particularly for bans on the use of chemicals. The international community, at the political level—that is, at the level on which states made treaties with each other—was close to unanimous in its abhorrence of weapons that relied on poisons and pathogens.

That abhorrence lives on to this day. In its public pronouncements, the global community overwhelmingly condemns the production and use—and even the stockpiling—of CBW armaments. Toward the end of the last century and into the new millennium, it has invested a good deal of time and energy in forging agreements aimed at slowing the proliferation of these weapons, with the professed aim of ultimately banning them altogether. Two agreements in particular, the Biological and Toxin Weapons Convention of 1972 (the BTWC) and the Chemical Weapons Convention of 1993 (the CWC), have been almost universally adopted by nations around the globe. Yet the world today has more nations than ever devoting their scientific skills and fortunes to developing CBW capabilities.

For a variety of reasons, the CWC and BTWC have not succeeded in eliminating the threat these weapons pose to international security. Some states, notably those in the Middle East, are not participating in the agreements. Egypt, Lebanon, Libya, Syria, and the United Arab Emirates, for example, have neither signed nor ratified the CWC, while Israel has signed but not ratified the agreement, and Iraq to this day remains non-compliant even though it agreed to abide by the terms after being compelled to do so at the end of the Gulf War. Moreover, some states, after signing and ratifying, have simply proceeded to violate the terms, the most notorious example being the former Soviet Union, which ratified the BTWC in 1975, then embarked on a massive expansion of and new investment in its programs of BW research, development, and production throughout the 1970s and 1980s—right up until it ceased to be a nation, and perhaps beyond.

Many in the international community are clearly unwilling to lay down these arms, their public professions of willingness notwithstanding. If one judges the treaties on how close they have come to completely eliminating CBW armaments, they have to be considered failures. On the other hand, if the treaties are judged on how few major CBW battles have been fought over the last few decades, then each can be seen, for now at least, as a qualified success.

The current CBW status of the countries discussed in this chapter may not, at first glance, inspire feelings of security and stability, but it does have to be said that over the last 25 years, and especially over the last 10, there has been an increased flow of information among nations that have ratified the conventions, particularly in relation to chemical weapons. Some antagonists may not be less inclined to develop and stockpile CBW armaments, but all nations are now certainly aware, in more than just a vague and general way, which states have, or are likely to have, or are soon to have, biological and chemical weapons. That is the provisionally good news. The bad news is that one cannot rely on a compact among nations—"State Parties," in the language of arms control—to definitively control or even keep track of those who wish to arm themselves with these weapons. At the end of this chapter, we briefly consider the matter of so-called "sub-state" entities. This cast of characters is by now well known—terrorist organizations, fanatical cults, rogue regimes, global crime syndicates, and guerilla warriors are all potential "actors," and although we put them in a separate category, it is not always clear who are their sponsors. There can be no doubt, however, that they would like to play a global role in how chemical and biological warfare is conducted, and so we will take some time to see, as best we can, how they fit into the picture.

THE SUPERPOWER AND FORMER SUPERPOWER

At the end of World War II, the United States and the Soviet Union proved to be the nations most capable of and willing to invest in chemical and biological weapons. Although most of each country's armament expenditures during the Cold War were directed at the development of nuclear capabilities, each amassed a huge stockpile of chemical weapons—in the case of the Soviet Union, something on the order of 40,000 tons, and with the US close behind with some 30,000 tons. By the 1970s, the USSR, lagging in the nuclear arms race, re-energized an already-large biological weapons program that in its sheer output probably overshadowed the BW capabilities of all other nations combined. What happened in the last two decades of the century, however, was that the leaders in CBW armaments, in very different ways, shaped and to a large extent directed the worldwide movements to control and ban CBW weapons.

The United States

In 1969, shortly after he took office, President Richard Nixon declared, "the United States unilaterally renounces first use of lethal or incapacitating chemical agents and weapons and unconditionally renounces all methods of biological warfare."[1] Following this declaration, the United States limited all research on chemical and biological weapons to developing defensive agents and antidotes. In 1975, the United States ratified the BTWC, and in 1993, the US took the additional step of signing and eventually ratifying (in 1997) the Chemical Weapons Convention, forswearing possession, development, and use of all chemical weapons, and committing itself to destroying its stockpiles of the weapons by the end of 2004. (The CWC itself requires all states that possess chemical weapons to destroy their stockpiles by April 2007.)

Chemical Weapons

The United States was a latecomer to World War I; it did not declare war on Germany until April 1917. By early September of that year, a "Gas Service" had been established as a separate branch of the American Expeditionary Force in France, but it was not until June 1918, five months before the Armistice, that members of the newly formed US Army Chemical Warfare Service, or CWS, became available for action on the front. Because of the risk of friendly-fire casualties and the fact that using CW drew a disproportionate amount of enemy fire, US Army field officers resisted engaging in gas warfare. It was later said by Major General William L. Sibert, the first commanding general of the CWS, that the service actually had to "go out and sell gas to the Army."[2] In the end, the US Army command did overcome its reluctance and

Major General William L. Sibert brought disparate elements of Gas Service into one Chemical Warfare Service. He also commanded the 1st Division in France in early 1918. (Courtesy of Soldier Biological and Chemical Command, Historical Research and Response Team, Aberdeen Proving Ground, MD.)

brought phosphorus smokes, bombs, and flame-generating ordnance, as well as poison gas, into battle, and the CWS "began to handle offensive gas operations in the way they should be handled."[3] While most of the mortars used in these attacks were phosphorus incendiaries, poisonous gas was included in some 200 actions by the American First Gas Regiment from the second battle of the Marne in July 1918 up until the end of the war. American soldiers were injured and died in gas attacks (suffering nearly 100,000 gas casualties), but these numbers were almost negligible when compared to the gas casualties inflicted on English, French, German, and especially Russian forces.

Nonetheless, at the League of Nations-sponsored World Disarmament Conference in 1932, the United States, as an observer to the proceedings, was prepared to declare that it would not be the first to use chemical weapons in any conflict, and to eliminate the use of all lethal gases in combat (some non-lethal gases excepted). In 1933, a British counterproposal went even further. Called the "MacDonald Plan," it proscribed *all* military preparation for gas warfare. At first this proposal seemed well-timed; Franklin D. Roosevelt, who had just been elected President, seemed inclined to accept the plan. However, by the time it was proffered for signature, Hitler had come to power, Japan had withdrawn from the League of Nations, and the World Disarmament Conference had fallen apart in disarray.

CW in World War II

In June 1943—reflecting the general disapproval of chemical warfare by the American public—Franklin D. Roosevelt categorically stated that "we shall under no circumstances resort to the use of such weapons unless they are first used by our enemies."[4] By 1945, however, the public mood had shifted somewhat, with 40 percent of those polled in favor of using chemicals against the Japanese when just a year before the number had been 23 percent.[5] The shift in opinion was in part the result of pitched battles in the Pacific islands, which had caused horrendous US casualties.

As plans were being drawn to invade the Japanese home islands, General "Vinegar Joe" Stilwell and General George C. Marshall suggested the use of gas. While President Harry Truman had yet to formalize policy on the subject, other influential decision makers within the military, notably Admiral William D. Leahy, thought it appalling that chemical weapons were being considered at all. In 1945, the United States could have mustered enough chemical weapons from existing stocks, and had the industrial capacity to manufacture many thousands of tons more. But without gas masks and other equipment forwardly deployed—not to mention the lack of CW training among US troops—practical considerations in addition to moral qualms mitigated against the use of CW. Moreover, those at the very highest level of government knew that, before long, the US might very well have an even more decisive weapon to use against the Japanese.

In the wake of World War II, several dozen committees and advisory groups working under the auspices of the United States Department of Defense (DOD) examined chemical weapons-related issues and recommended, in light of the military threat from the Soviet Union, that the US needed to expand CW production capabilities, research, and public-relations efforts. The CWS, which at this point had been part of the US Army for almost 30 years, was finally renamed the Chemical Corps in 1946. Because it had never really played a central role in US military operations, the Corps periodically had to overcome widespread anti-CW sentiments, often from within the US Army itself. It did, though, manage to survive, largely through a combination of support from key legislators, the civilian chemical industry, and vigorous public-relations campaigns.

CW During the Cold War

On June 30, 1950, the US DOD *ad hoc* Committee on Chemical, Biological and Radiological Warfare issued the Stevenson Report, which noted that, in the aftermath of World War II, the Soviet Union had acquired two German CW production plants, one for the production of sarin and the other for tabun, while the US did not yet have any comparable large-scale facilities. The report estimated that it would take the US approximately two years to secure a significant production capacity. It recommended that the "necessary steps be taken to make the United States capable of effectively employing toxic chemical agents at the onset of war."[6] Although the report was ultimately withdrawn, its recommendations were influential in shifting the policy debate. Following the outbreak of the Korean War, the US Congress finally approved the funds to construct a sarin production facility as part of an emergency military appropriations bill.

In 1953, a $50-million facility was constructed in Muscle Shoals, Alabama, to supply precursors for nerve agent production at the Rocky Mountain Arsenal in Denver, Colorado. The final products were G-series agents, sarin and soman, while other CW agents were also being manufactured at Edgewood Arsenal, the US Army site in Maryland that had been producing toxic agents and chemical munitions since 1917. By 1962, sarin (GB), VX, CS tear gas, and BZ (an incapacitating agent) had become standardized components in the US military's offensive CW arsenal.

In the United States, as elsewhere, the chemical weapons industry was tied closely to industries that manufactured pesticides, insecticides, and fertilizers, all of which relied heavily on research on organophosphorus compounds—that is, substances containing phosphorus and carbon elements in their struc-

Filling of 75-mm artillery rounds with mustard. Note the lack of protective equipment for the personnel. (Courtesy of Soldier Biological and Chemical Command, Historical Research and Response Team, Aberdeen Proving Ground, MD.)

ture. Especially in the course of synthesizing and testing potential pesticides—many of these based on organophosphorus compounds—chemists frequently came upon substances that were too toxic to animals to be commercially viable, but were suitable for additional testing as potential chemical weapons.

DDT is an organochlorine insecticide; it was truly a "miracle" compound, and is credited with saving millions of lives from mosquito-borne diseases. By 1952, however, it was losing its effectiveness as insects developed resistance, so scientists worldwide were doing organophosphorus insecticide development in an attempt to find a suitable replacement. Dr. Ranajit Ghosh at the British firm Imperial Chemical Industries was conducting pesticide research when he synthesized an exceptionally toxic nerve agent that killed animals as well as insects. Seizing upon its possible military significance, he handed over the formula and sample of this organophosphorus compound to the British government defense laboratories at Porton Down in England. However, the British military establishment had already decided to adopt one of the G-series nerve agents, and had begun building a chemical arsenal that used either tabun (GA) or sarin (GB). Having already committed to this program, the United Kingdom decided to give the formula to the US and Canada.

American chemists at the Edgewood Arsenal laboratories made structural changes to Ghosh's original formula and coded it as VX. This viscous liquid, a highly toxic nerve agent, became a staple of the American chemical arsenal. (In 1955, the Soviet GRU, or military intelligence, managed to steal Ghosh's structural formula for V-agents. Ultimately, the USSR developed its own version of the agent, V-gas, which differs slightly from VX but shares similar physical, chemical, and toxic properties.) Production of VX and other nerve agents continued through the 1960s—until there were two unsettling incidents involving them: one in Skull Valley, Utah, and another in Okinawa, Japan, both occuring in 1969. In the first instance, an exercise involving the delivery of VX aerial munitions near Dugway Proving Grounds, Utah, went awry, resulting in the accidental release of about 20 pounds of the agent. No humans were injured, but 6000 sheep died in what became a highly publicized and very embarrassing accident. Later that same year, 23 American soldiers and a US civilian were exposed to sarin while refurbishing chemical munitions. The public outcry in Japan helped hasten the removal of chemical weapons from US installations in the Pacific, as well spur President Nixon to renounce first use of them in any war. As stated earlier, research and development of offensive CW armaments for all intents and purposes came to a halt.

That is not to say, however, that researchers in the Chemical Corps were idle from 1969 onward. In fact, their work expanded into the development of napalm-like incendiaries (used in Korea and Vietnam), herbicides, riot-control

The Honest John warhead held numerous bomblets filled with sarin nerve agent. It was designed to scatter the submunitions near the target. (Courtesy of Soldier Biological and Chemical Command, Historical Research and Response Team, Aberdeen Proving Ground, MD.)

agents, and even "people sniffers"—devices that detected human odors. They also produced Agent Orange, the controversial herbicide used in Vietnam.[7]

In 1985, the administration of President Ronald Reagan ended the US's unilateral 17-year moratorium on offensive CW development and production. The Department of Defense invested significant amounts of money in developing binary chemical weapons, including some extremely sophisticated devices in which the mixing of the components took place "in flight"—that is, while the shell or bomb or rocket was on its way to a target. The US Navy also restarted its "Bigeye" program, featuring an aerial munition containing separate compartments of sulfur and QL, the compound representing the nearly complete molecule, that were mixed together during the shell's flight to the target, creating the deadly nerve agent VX. Because of technical difficulties, skepticism in Congress, and sensitive arms negotiations with the Soviet Union, this program was eventually abandoned.

However, binary production was not stopped entirely. On December 16, 1987, a 155-mm binary nerve artillery shell went into production. This munition held the precursors DF (difluor, short for methylphosphonic difluoride) and isopropyl alcohol, which by US law were required not only to be stored in different depots, but in different *states*. After firing, discs separating the precursors would rupture from the force of sudden acceleration (about 8 Gs), while the fast spin of the projectile (15,000 rpm) would yield a 70-percent mixture of sarin.

Current Status of US Chemical Weapons

Today, the United States holds the second-largest chemical stockpile in the world, of approximately 28,000 agent tons. All stores of chemical munitions, primarily blister and nerve agents, have been slated for complete destruction by mandate of the CWC. Because of some technical difficulties and environmental concerns, whether or not these chemical stores will be completely destroyed by 2004 is still uncertain, but destruction is proceeding.

Although research in offensive CW has been discontinued in the United States, particularly since the Gulf War (1991), the US military has steadily increased spending in the field of chemical defense. The vast majority of developmental funds and energies goes to helping protect and sustain the American soldier, with much of the research and development being conducted at the US Army Research Institute for Chemical Defense. Among the most pressing needs in CW defense are those involving accurate and dependable detectors, improvements in protective garments, and drugs for the treatment of chemical casualties.

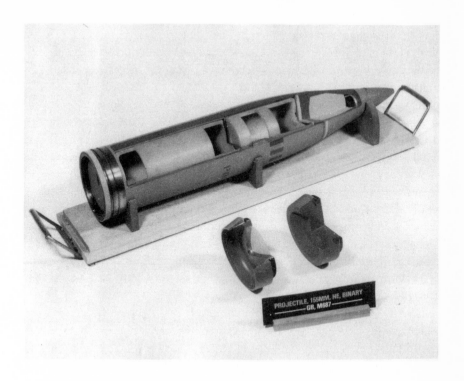

Although standardized in 1976, the US military only produced the M687 binary (sarin) 155-mm projectile in the late 1980s. (Courtesy of Soldier Biological and Chemical Command, Historical Research and Response Team, Aberdeen Proving Ground, MD.)

Biological Weapons

Just prior to World War II, the United States received domestic intelligence reports indicating that Japan and Germany had undertaken research into biological weapons. These reports and other intelligence created a sense of urgency at the War Department, and in September 1939 American military scientists decided to examine the problem of BW, which as an issue had mostly lain dormant for almost 15 years. (Biological warfare was a collateral issue during an earlier disarmament conference, the 1925 Geneva Protocol, in which the US was a participant.) The diseases that concerned the US government included yellow fever, dysentery, cholera, typhus, bubonic plague, smallpox, influenza, and sleeping sickness, all of which could be spread by insects. In earlier discussions, it was held by the Chemical Corps and most in the War Department that diseases like these could have devastating impact on a battlefield, but it was also believed that strict sanitation and public health practices (for example, eliminating mosquitoes, lice, and rodents) were sufficient to safeguard troops.

Three years later, faith in the public-health approach to BW defense was somewhat less secure. In a May 1942 cabinet meeting, President Roosevelt, at the urging of Secretary of War Henry L. Stimson, established the War Research Service (WRS) in order to oversee issues related to biological warfare. George W. Merck, President of Merck & Co., the huge pharmaceutical concern, was made chairman and given the responsibility of building and operating laboratories and production facilities. In a 1942 WRS report, anthrax, tularemia, and agricultural agents like foot-and-mouth disease and potato blight were listed as urgent threats, but also potential weapons for offensive use.

From 1943 to 1944, the US created domestic BW testing facilities at Camp Detrick (later Fort Detrick) in Maryland, the Dugway Proving Grounds in Utah, and Horn Island in Mississippi, among others. The US also undertook joint research and development projects with Canada and Britain. At outposts in Scotland and Wales, for example, anthrax bomblets included in cluster munitions were tested on sheep and found to be highly effective. Though many of these ventures yielded agents and delivery systems that could, conceivably, have been used in the war against Germany and Japan, none of them were used in combat.

The Post-World War II Era and the Korean War

Immediately following World War II, production of biological agents in the United States was essentially reduced from factory-level to laboratory-level output. However, work on delivery systems—the technology of getting agents to the target—was markedly increased. During the Korean War, the United

States established the Pine Bluff Arsenal in Arkansas for BW production and testing, and enlarged the BW research facilities at Fort Detrick. At these and other facilities, the Department of Defense conducted small-scale open-air testing of the delivery of *simulants*—ordinary bacteria or chemical markers that in their physical behavior as airborne particles mimic full-fledged BW agents. The US government firmly maintained, however, that BW weapons were never used in Korea, in spite of accusations made by the North Koreans, Chinese, and Soviets. The Chemical Corps and US Army and Air Force made significant advances in its BW capacities at this time, particularly in terms of delivery systems. Tests to determine the effectiveness of mosquitoes as delivery agents, tests in which simulant bacteria were released over US cities, and tests with human volunteers were among the experiments conducted covertly during this period.

Despite increasing public interest in disarmament during the 1960s, the American offensive BW program continued to grow. By 1966, the facilities at Pine Bluff and Fort Detrick had already mass-produced several BW agents for use in a variety of munitions. By the time its BW activities came to an end in 1969, the US had seven standardized biological weapons: the lethal bacterial agents that cause anthrax and tularemia, the agents used to create the incapacitants brucellosis, Q-Fever, and VEE, and the lethal toxin botulinum. SEB was also produced as an incapacitating toxin.[8]

By 1972, following President Nixon's National Security directives renouncing all offensive development and production of microbial and toxin agents, all anti-personnel BW agent stocks and munitions were destroyed. The United States also terminated all offensive research and closed or cleaned up all offensive facilities, turning them over to other government agencies for research. The unilateral disarmament initiated by President Nixon's directives in a sense set the stage for the 1972 Biological and Toxins Weapons Convention (BTWC), which was finally ratified on January 22, 1975.

Russia and the Former Soviet Union

Even after the fall of the Soviet Union in 1991, the Russian federation of states had in place what was probably the largest chemical and biological warfare infrastructure in the world. Not only were there massive stockpiles of chemical armaments, but tens of thousands of troops, thousands of scientists, and hundreds of research and production facilities spread across at least six vast and now autonomous states. For the world community, the problem was that Russia's 75-year history of CBW research, development, and production had created a legacy not just of munitions, but of a highly skilled and experienced corps of scientists and soldiers who no longer worked under strict central control, and who for all intents and purposes were now actually unemployed. The remains of the clandestine Soviet CBW programs was now not just Russia's problem, but the world's.

Chemical Weapons

Since the end of World War I—and especially throughout the Cold War—the Soviet Union maintained a highly capable and formidable CW capability. At its height, the Soviet army had 80,000 CW troops, a number that could have easily doubled with reserves in wartime.[9] The modern Soviet chemical arsenal contained approximately 40,000 tons of CW. Nearly all Russian artillery systems had some chemical-weapons capability, as did mines and aerial spray tanks. But the beginnings of Soviet chemical weaponry were less than auspicious.

Russia's armies suffered nearly 500,000 CW casualties during World War I, a toll that amounted to 62 percent of all chemical casualties suffered by all sides throughout the war. On July 24, 1916, Russian troops used gas offensively for the first time against German troops at Skrobsk, and as related by Alexander Solzhenitsyn, the effort had disastrous consequences:

> . . . This summer one regiment was planning to use gas—three
> emissions of a hundred canisters, beginning at midnight—and
> then attack. But they dilly-dallied too long and released the first
> wave only at 3 A.M. The Germans detected it—rockets went up,
> trumpets and horns sounded the warning, iron sheets were ham-
> mered, beacons were lit. Our meteorological station then reported
> that the wind was becoming changeable, but the division com-
> mander ordered the release of the second wave. Men in the
> neighboring regiment, in a slightly forward position, were gassed.
> The wind became less favorable—but the third wave was ordered.
> This one traveled a little way, stopped, and was blown onto our
> trenches. To make things worse—the canisters should have been
> placed in front of the trenches, with their pipes pointing toward
> the enemy, instead of which, contrary to instructions, the canisters
> were left in the trenches, with their pipes resting on the parapets.
> The Germans opened fire on our trenches, smashed the canisters,
> and panic-stricken men had to tug on gas masks in a hurry. Three
> hundred of them, officers and men, were buried in a common grave.
> The division commander's suspension was a poor consolation. . . .

> —*Alexander Solzhenitsyn, November 1916*[10]

With the World War I catastrophe etched in the minds of Red Army officers, the Soviet Union embarked on an aggressive chemical weapons development program, fully integrating CW doctrine within its armed forces.

Although it was clear that an adequate technical base had yet to be built, the Soviet doctrinal marriage of chemical weapons in warfare had already been developed by 1920. In January of that year, instruction in CW defense was formalized in the curriculum of the Higher Military-Chemical School of the Red Army.[11]

Although Western countries participated in the nascent civilian industry of Soviet Russia, it was a marriage of convenience with Germany that fostered much of the military chemical manufacturing for both sides. By partnering with the Germans, Russia was able to produce large quantities of CW agents. After Hitler seized power in 1933, however, the cooperation ceased.

At the end of World War II, Stalin was able to acquire at least two large German CW factories relatively intact. One of them, the Dyenfurth tabun production plant, was taken apart by the Soviets and reassembled in Volgo-grad.[12] The Soviets aggressively studied analogues of sarin and soman and investigated as well an early prototype of a nerve-agent detector.

According to Major General N. S. Antonov, former commander of the Military Chemical Establishment at Shikhany, the Soviets first received detailed information about the production of nerve gas in 1957. Ghosh's for-mula for VX nerve agent inspired a series of experiments that finally led to the development of Russian V-*gaz*. Around the same time, the institute changed its name to the Central Scientific Research Military-Technical Institute (TsNIVTI), and later moved from Moscow to Shikhany, combining it organi-zationally with the Red Army Central Military Chemical Proving Grounds (TsVKhP).[13]

The Soviet CW program was not strictly limited to domestic stockpiling. When Egyptian weapons were captured during the 1973 Arab-Israeli War, the Israelis discovered that the munitions were Soviet-made and that the armored vehicles, especially tanks, had built-in nuclear, biological, and chemical filtra-tion systems. Egyptian soldiers had also been equipped with nerve agent anti-dote kits, including preparations made specially for treating soman casualties.[14] Although no actual chemical armaments were found, these discoveries, which came as something of a surprise to US intelligence, were alarming because they suggested great depth and breadth in the Soviet Union's CW capacities.

The Russian Federation in the 1990s

As a result of ongoing bilateral agreements with the United States and legisla-tive action taken by the Russian parliament, or *Duma*, Russia is beginning to destroy its CW agent arsenals. Although a few Russian politicians resisted the idea entirely, former President Boris Yeltsin signed the "Comprehensive Destruction Act," which called for the destruction of Russia's CW agent stockpiles in 1997. However, Russia lacked the estimated $5.7 billion needed to do the job, so the completion of this project will depend on the international aid Russia receives. There is great concern that the project will not begin quickly enough. As Colonel-General Stanislav Petrov, chief of the Radiological, Chemical, and Biological Protection Troops, warned, the longer Russia waited to destroy its CW agent arsenals the more likely terrorists or "madmen" could steal them. A State Party to the CWC, Russia is obligated to destroy its massive stockpile, which as mentioned is estimated at 40,000 tons. However, there appears to be an understanding that Russia's economic woes are certain to cause further delay in its destruction. Some of the remaining stockpile has been slated for industrial use. In an interesting effort to recycle and recover costs, large stockpiles of Lewisite will be reduced to 7000 tons of arsenic metal that can be utilized for gallium-arsenide semiconductor production. Although the cost-effectiveness of this program is still in question, it is hoped that this arsenic would help foster industry in Russia.[15]

Biological Weapons

Much that is known about the former Soviet Union's BW capacities was revealed by two defectors: Vladimir Pasechnik, the Soviet defector to England who revealed the BW activites of Biopreparat in 1989, and Kanatjan Alibekov (now Ken Alibek), who has given the West more detailed information about biological weapons research. From these important sources and others, William C. Patrick III—who developed offensive biological research for the United States—has surmised that BW research and development in the USSR "paralleled ours very closely."[16]

In 1989, after Pasechnik was in the hands of the British government, both the United Kingdom and the United States pointedly asked Soviet President Mikhail Gorbachev about the Russian BW program. Gorbachev adamantly denied its very existence. In January 1992, however, Russian President Boris Yeltsin admitted that the former Soviet Union had experienced a "lag in implementing" the BTWC.[17] The following year, after a visit by Yeltsin to the US and further intelligence work, the United States released this assessment of the situation:

> ... The United States has determined that the Russian offensive biological warfare program, inherited from the Soviet Union, violated the Biological Weapons Convention through at least March 1992. The Soviet offensive BW program was massive, and included production, weaponization, and stockpiling. The status of the program since that time remains unclear. ...[18]

Why had the Soviets failed to disarm? According to Alibek, when President Nixon renounced all forms of biological and toxin warfare in November 1969, Soviet leaders doubted his sincerity, and they "strongly believed that the United States had an offensive program."[19] The thinking inside the Soviet Union, according to Alibek, was that with the United States continuing its offensive research, the USSR had no choice but to continue as well. And proceed they did: The microbes considered most suitable for weaponization included smallpox virus, anthrax, and plague bacteria, but the Soviet military planners and scientists also studied some 50 other biological agents.[20] Most startling were Alibek's revelations about the immense scale of the smallpox program, and the fact that tons of smallpox agent were weaponized for delivery in intercontinental ballistic missiles.[21]

The turnaround in Russia, when it came, was emphatic, although there is disagreement on how closely the edict to disarm its BW program is being followed and monitored. According to an April 1992 Soviet decree, any and all biotechnology research that ran counter to the BTWC was banned in Russia. As a consequence of this policy—as well as economic considerations for privatizing government-run industries—former BW-related entities are beginning to produce items for public health and agriculture. Despite resistance among military and nationalistic forces, the Russian political leadership appears to have been sincere in committing itself to the BTWC. Unfortunately, the conversion of former BW defensive- and offensive-related facilities to fully commercial enterprises in the former Soviet Union has been met with only limited success. And as we have learned, compliance promised does not necessarily mean compliance occurs. During an interview with *Prime Time Live*, Ken Alibek said research continues on biological agents in the newly-formed Russian Federation.[22] Alibek also argued that at least four major bioresearch centers of the former Soviet BW apparatus—still off-limits to outsiders—must be thoroughly inspected for ongoing BW activities.

As with the Russian CW program, the many scientists that have previously worked in BW research and are now unemployed create a fear of "brain drain." In 1995, a number of reports suggested that Iranian advances in BW research were aided by former Soviet scientists, and similar reports have suggested that Iraq and other countries may have made offers as well. Another report tells of a former Russian BW scientist who offered his services to China. In an effort to curtail these developments, Western-financed initiatives such as the International Science and Technology Center (ISTC) in Moscow work at directing Russian weapons scientists toward work on more peaceful projects.

THE MIDDLE EAST

The Middle East is perhaps the most volatile hot-spot in the world and has commanded much of America's diplomatic and military attention since the late 1960s. The series of full-scale wars between Israel and its neighbors have, of course, been a predominant concern, but so too has the development of numerous terrorist organizations, some of them sponsored by Middle Eastern states.

Iraq

Iraq demonstrated its CW capabilities during the horrific 1980–88 Iran-Iraq War. In the years leading up to the Gulf War, the Iraqi government also began to develop a substantial biological weapons capacity. In fact, when Iraq invaded Kuwait, they had the most extensive biological, and possibly chemical, arsenal in the Arab world, reflecting a history of more than ten years of research and development.

Chemical Weapons

Iraq's initial acquisition of chemical weapons had a two-fold purpose. First, it was meant to achieve clear military objectives during the 1980–1988 Iran-Iraq War. Second, the Iraqis hoped that a CW arsenal would deter other states from interfering with Saddam Hussein's adventurism in the Persian Gulf. More specifically, Iraq believed their chemical arsenal would allow them to threaten neighbors and offset Israel's nuclear capability by deterring preemptive strikes on Iraqi nuclear facilities. Finally, Saddam Hussein intended to use CW as part of an active program to depopulate anti-government Kurdish elements, especially those within Iraqi borders, even if that meant simply killing civilians.

Iraq produced and weaponized large quantities of blister agents (mustard, Lewisite) and nerve agents. The liquid mustard agent that Iraq produced was of good purity and had a long shelf life. Iraq also produced "dusty" mustard, made by absorbing the liquid agent onto a talcum-like powder. Dusty mustard is easier to disseminate as a breathable aerosol and is more concentrated.

As it built its chemical arsenal, Iraq also manufactured hundreds of tons of nerve agents, including tabun, sarin, cyclosarin (GF), and VX. Iraqi tabun and sarin were of poor quality, containing impurities that gave them a shelf life of only 4 to 6 weeks. However, there is evidence that Iraq may have produced hydrogen cyanide gas with an additive to make this blood agent more persistent.

The Gulf War (1990–1991)

Chemical weapons also figured into Iraq's war plans as it braced for a show-down against US-led Coalition forces just prior to the Gulf War. The Iraqi

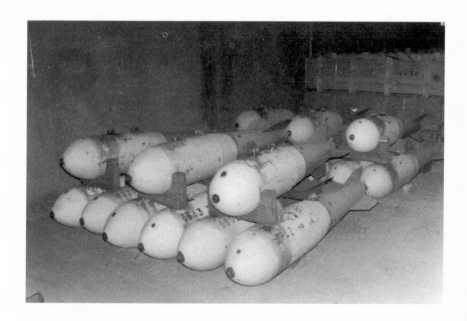

R-400 nerve agent aerial bombs found at Iraqi Al-Walid airbase. Iraq also filled this type of ordnance with Bacillus anthracis *(anthrax) spores. This photograph was taken in fall 1991.* (Courtesy of United Nations, Photograph by H. Arvidsson.)

military prepared its CW apparatus near the Kuwaiti Theater of Operations, and according to early intelligence reports from Iraqi commanders, chemical weapons were also in place within Kuwait. Iraqi plans included the use of chemical weapons to deter attacks by raising the prospect of mass casualties, a perceived weak spot among Coalition forces, especially for the United States. Available evidence indicated that the Iraqi leadership intended to employ CW weapons on the battlefield, and that unit commanders had been trained specifically to use them. In September 1990, the US Defense Intelligence Agency was confident that, in addition to having munitions that could deliver chemicals, Iraq possessed tabun (GA), sarin (GB), cyclosarin (GF), mustard, and CS (tear gas).[23]

During the build-up for Operation Desert Storm, the United States realized that it was woefully unprepared for an all-out chemical war with Iraq. At the time, the US government contemplated moving stored chemical munitions from Germany in the event that chemical retribution was required. However, this idea was abandoned rather quickly, and a two-pronged campaign was put in its place. First, it was decided that a media blitz using news outlets such as CNN—no doubt watched constantly by the Iraqi leadership—would showcase US soldiers fitting up with the latest in chemical and biological warfare defense garments and detection equipment. In an attempt to demonstrate the Coalition's readiness for CW, additional airing of press releases showed the American troops training for chemical attacks, even though privately US military leaders were quite aware that no significant training program had taken place.

Secondly, not-so-subtle hints were made that nuclear weapons might be used against Iraq if chemical weapons were launched against Coalition forces.[24] President Bush, while not explicitly mentioning the nuclear option, did threaten massive retaliation in a January 1990 letter to Saddam Hussein:

> . . . Let me state, too, that the United States will not tolerate the
> use of chemical or biological weapons or the destruction of
> Kuwait's oil fields and installations. Further, you will be held
> directly responsible for terrorist actions against any member of
> the coalition. You and your country will pay a terrible price if
> you order unconscionable acts of this sort. . . .[25]

To make ready for such a contingency, a "Punishment Air Tasking Order" was drawn up by the US Air Force detailing the targets that would be destroyed in the event of a chemical attack. In the end, no chemical weapons are known to have been use by Iraq, and although fear of nuclear retribution appears to have been key, the following factors also probably played a role:

• Iraq's belief that Coalition forces were prepared for chemical weapons attacks (even if this level of preparation was exaggerated).

· The prevailing wind was no longer blowing south, posing a risk to Iraqi troops from their own chemical weapons.[26]

· Inclement weather.

· The disruption of the Iraqi command-and-control network as a result of the Coalition's air campaign.

· The rapid advance of the Coalition's ground forces.

Iraqi VX

As a condition of Iraq's cease-fire agreement in 1991, the United Nations Special Commission (UNSCOM) was established and given the task of finding any remaining Iraqi weapons of mass destruction, including any and all chemical munitions. In 1996, after a good deal of evasion, Iraq finally told UNSCOM inspectors that it had built a VX nerve agent manufacturing facility in Muthanna. The Iraqi government also claimed that it had unilaterally disposed of nearly 4 tons of VX. During an UNSCOM verification process, samples of missile fragments collected from the dumpsite were found to contain VX degradation products, indicating that VX had indeed been dumped there.

Further chemical analysis conducted by the US Army in Aberdeen, Maryland, detected two compounds that clearly showed VX degradation products—the remains of the agent. While there was no doubt that VX had been produced by Iraq in large quantities—if only because Iraq had admitted as much—the remaining point of contention was whether Iraq had loaded its "special missile warheads" with the agent. Ten of the eleven missile fragments analyzed in Aberdeen yielded evidence that suggested the presence of VX. The remaining piece tested showed traces of methylphosphonic acid, a sign of either V-agent or G-agent (such as sarin). These results were given to UNSCOM in June 1998, and the Iraqi authorities were confronted with the evidence during the middle of that same month. They rejected the findings and even charged that someone—presumably the United States—had planted VX on the samples on the way to the testing laboratories. Iraqi Deputy Foreign Minister Al-Qaisi insisted that "The incontrovertible facts are that Iraq never produced VX in stable form and never filled VX in warheads," and the VX degradation product findings were "the result of a deliberate act of tampering with the first set of samples taken out from Iraq to the United States."[27]

Inspections stopped in 1998, when the Iraqi government stopped cooperating and essentially threw the inspectors out of the country. As of the fall of 2001, the United Nations Security Council has yet to persuade Baghdad to restart inspections, agreement to which is necessary to remove economic sanctions on Iraq.

Even without the inspections, the US has been able to gather more information about Iraq's CW production. According to a former head of Iraqi military intelligence, General Wafiq al-Sammarai, who defected to the west in 1994, VX production began as early as May 1985 and continued until December 1990 on an industrial scale. To keep the VX stable over long periods, it was made into a salt and stored until it was needed, at which point it would be reconstituted with an alkaline substance similar to baking soda.

General Sammarai further informed the West that Iraq gained the capability to use VX on the battlefield much earlier than previously thought. For example, Iraqi forces may have used nerve agents against Iranian forces to great effect on the Al Fao peninsula in 1988.[28] According to another report, at least six months prior to this battle, German scientists were recruited by the Iraqi government to assist in the weaponization of VX.

Biological Weapons

> . . . Iraq claims that all biological agents and munitions were uni-
> laterally destroyed after the Gulf War. However, Iraq's record of
> misrepresentation and the lack of documentation to support these
> claims leave the status of Iraqi biological warfare stockpile in
> doubt. . . .
>
> —US Department of Defense, 1996[29]

Although Iraq signed the BTWC in 1972, the government only ratified the treaty in 1991 as a condition of the Gulf War cease-fire. The information gathered through UN weapons inspections following the Gulf War showed that Iraq worked on BW at facilities in the towns of Salman Pak and Al Hakam. There, Iraq mass-produced anthrax, botulinum toxin, ricin, and aflatoxin, and investigated the use of many other potential BW agents.

We now know that from 1974 to 1978, the Iraqi government instituted a biological weapons program, but did not begin research and development in this area until 1985. After receiving bacterial strains from abroad in April 1986, practical work in the growth of BW agents and their weaponization restarted. In 1990, the Iraqis claimed that over 6.5 tons of botulinum toxin and over 9 tons of anthrax were cultured at Al Hakam. At about the same time, Iraq's Daura Foot-and-Mouth Disease Institute produced another 6 tons of botulinum toxin, plus another 400 liters at Taji. An estimated 150 liters of anthrax were also produced at Salman Pak. Large-scale fermentation of *Clostridium perfringens* (gas gangrene) was also conducted, leading to the production of at least 340 liters of the organism and its toxins.

Using live animals on some occasions, Iraq conducted weaponization tests with simulants in early 1988. In November 1989, 122-mm rockets were

Al-Hakam BW facility, Iraq: UNSCOM inspectors installing monitoring cameras, 1995. This site was the primary BW agent production plant in Iraq. Note the large fermenter on the left. (Courtesy of United Nations, Photograph by H. Arvidsson.)

tested, with live fire experiments being performed using actual BW agents in May 1990. Starting in December 1990, more than 150 aerial bombs and 50 warheads were weaponized using BW agents, including a spray tank capable of holding up to 2000 liters (approximately 2 tons) of anthrax. While Iraq claimed tests with the spray tank were a failure, three more of these tanks were modified and placed in the Iraqi arsenal.

Another defector provided the West with a more complete picture of Iraq's BW development. After Lieutenant General Hussein Kamel left for the west in 1995, the Iraqi government admitted that it conducted an extensive BW program, but claimed to have destroyed all of its stockpiles. However, UNSCOM was unable to verify Iraq's claims.

Iran

Although the Iranian government has signed and ratified both the CWC and the BTWC, evidence suggests that they do have a chemical weapons arsenal and have at least attempted to build a biological counterpart.

Chemical Weapons

After unsuccessfully trying to mobilize world opinion against Iraqi use of chemical weapons in the 1980–1988 war, Iran launched a crash program to acquire its own CW defensive gear and chemical agents. The Islamic Revolutionary Guards, with support from the Ministry of Defense, were ordered to develop a full-fledged CW program.

In spite of these efforts, Iran had considerable difficulties with both defensive and offensive aspects of CW throughout the conflict. These problems stemmed from shortages of defensive gear and CW munitions and a lack of discipline and organization. As a result, the Iranian military could not respond effectively to Iraqi chemical attacks. However, using captured Iraqi stocks, Iran began using chemical weapons offensively between 1984 and 1985, and went on to use indigenously produced chemical weapons, including mustard and phosgene gas, between 1987 and 1988.

Even while supplying intelligence to Iraq during its war with Iran, the United States put export controls on chemical weapons precursors for export to both countries by 1986. Though largely playing a game of catch-up in the beginning, Iran continued to develop its CW program after the 1988 cease-fire, largely unhindered by the US export controls. Iran is currently suspected of having stockpiled a variety of CW agents and delivery munitions, including mustard, hydrogen cyanide, and phosgene, as well as sarin, tabun, and V-type nerve agents. Iranian chemical weapons could be employed on short- and long-range delivery systems, including the Shahin 2 rocket (20 kilometers) and Scud-type missiles (300–800 kilometers). To date, Tehran has added a wider array of agents in great quantities, with short-range rockets and longer-range missiles capable of delivering chemical weapons. Less certain are the organizational and operational aspects of Iran's CW capability. Further complicating the picture, Iran is a full-fledged participant in the CWC. However, to date no State Party has brought a "challenge inspection" request that would presumably clarify whether Iran possesses chemical weaponry.

Biological Weapons

While there is little direct evidence of a biological weapons arsenal in Iran, there is some evidence that its government has explored the option. First, according to the Henry L. Stimson Center:

> . . . The Iranian biological weapons program has been embedded within Iran's extensive biotechnology and pharmaceutical industries so as to obscure its activities. The Iranian military has used medical, education, and scientific research organizations for many aspects of biological agent procurement, research, and production. The US finding is that Iran probably has produced biological agents and apparently has weaponized a small quantity of those agents. . . .[30]

Furthermore, as the Stimson Center has also noted, Iran probably has both the technology and the infrastructure to support a large BW program, but even today there is insufficient unclassified data to accurately get a sense of the scale of its programs.

Syria

In 1967, Syrian Minister of Defense Hafez al-Asad was personally shaken by his country's disastrous defeat at the hands of Israeli forces during the Six-Day War. When he seized power as president four years later, Asad was determined to build up Syria's strategic capabilities. His dogged pursuit of weapons of mass destruction continued even if it meant that his conventional forces remained lackluster.

Chemical Weapons

Perhaps as early as 1972, Syria began the acquisition of what has now become a formidable CW capability. According to one estimate, Syria had an arsenal of both blister and nerve agents as early as 1986. While sarin has dominated most of Syria's stockpiles, they have also reportedly begun VX production. By the 1990s, Syria not only possessed hundreds of tons of CW agents, but also had deliverable chemical ordnance such as aerial bombs, artillery, older rockets, and perhaps even modern missiles similar to the Soviet SS-21.

It is difficult to say to what extent Syria relies on foreign expertise for chemical weapons development. Syria has received technical advice from the former Soviet Union and developed its more advanced chemical warheads with assistance also from North Korea and Western European nations. Reportedly, Syria is also able to produce CW agents indigenously. According to Major General Moshe Ya'alon, former intelligence chief of the Israeli Defense Forces, "Syria itself manufactures Scud B's, Scud C's, and chemical warheads of various types."[31]

In terms of the chemicals themselves, Syria does have a certain indigenous production capacity. A facility near Homs in western Syria may be the source of indigenous petrochemical derivatives for ethylene (precursor for mustard) and alcohol for nerve agents. Precursors for Syrian-produced CW agents have also been obtained from India and countries in Western Europe—or at least this has been the case in the past. While Syria's inchoate chemical industry leaves it dependent on foreign pesticides, it does have an indigenous source of phosphates, producing about a fifth of all phosphate rock mined in the Middle East. Phosphorus is an important element in nerve agent production, but it is uncertain how much (if any) is diverted to making chemical weapons in Syria.

The Syrian Center for Scientific Studies and Research near Damascus acts as a hub for CW research. Additional R&D is carried out covertly through a network of nationalized pharmaceutical firms. Hundreds of chemical weapons, including nerve and mustard gases, are produced near Damascus and Homs. Another chemical warhead production facility has been reported at Aleppo.

Because Syria possesses a combination of volatile (sarin) and more persistent (mustard, VX) CW agents, it is able to utilize chemical weapons in different tactical scenarios. Sarin is extremely deadly, but it evaporates at a rate similar to that of water. An attack using this agent would be ideal for creating large casualties closer to the front, but would dissipate quickly, allowing the aggressor to seize territory without great risk to his advancing troops. For example, an attack to retake the Golan Heights might include the use of such an agent, although Israeli troops on the front lines are well prepared for chemical combat. VX might be directed at targets behind the front lines, and at military installations and logistics. Because this nerve agent remains hazardous for a very long time, military energies would be diverted to the intensive decontamination procedures required. (One drop of VX on the skin can be lethal without fast medical intervention.[32])

Israeli-based publications and western analysts seem to be in agreement that Syria's Scud-C missiles, originally purchased from North Korea, are being utilized for long-range chemical weapons delivery. Recent reports indicate that Syria has also developed clusters of bomblets containing CW to be loaded into the 884-mm Scud-C warhead. Approximately 200 Syrian Scud-B missiles can be armed with VX and other unitary nerve (sarin) payloads. While the load capacity of Scud-Bs is superior to that of the Scud-C (985 kilograms and about 500 kilograms, respectively), the B's range is only 300 kilometers. Furthermore, the better accuracy of Scud-B missiles make them better suited for direct attack against Israeli military targets, meaning that the C might be used for attacks on more general, widely-targeted areas where Israeli reserve forces would gather in the event of a national emergency. The long range of the Scud-C gives Syria the capacity to hit most (if not all) significant targets in Israel—military, civilian, or otherwise.

Some Israeli and Western sources justify the conclusion that the Syrian Scud-Cs are armed with chemical warheads based on the following reasons:

- The Scud-C has long range but is very inaccurate, making it more likely a weapon to be used against populated urban centers.
- Syrian Scud-Cs deployed near Hama are supplied with a high ratio of launchers to missiles, enabling Syria to launch many salvos at once. This reflects a doctrine that allows Syria to strike decisively before the Israelis respond with nuclear weapons.
- Multiple missile firings can inundate Israel's present and perhaps even future theater missile defense.
- Specially hardened bunkers and tunnels have been constructed to protect Scud-Cs deployment, indicating their strategic value.

Because much of Iraq's chemical weapon arsenal has been found and destroyed by UNSCOM, Syria may now dominate the region in terms of CW agent production and delivery systems. Conventional wisdom suggests that Syria uses its chemical arsenal, especially sarin and VX nerve agents, to deter a possible Israeli nuclear strike, or, at the very least, to be able to launch a devastating first strike before Israel has a chance to respond with its own nuclear weapons.

Biological Weapons

Given Syria's substantial CW arsenal, the US suspects that they may have developed a biological weapons stockpile as well. Israeli intelligence believes that the Syrians have developed BW agents to contaminate drinking water, but the evidence, at this point, is inconclusive.[33] Other evidence suggests that Syria has not yet moved beyond the research and development phase and probably has only produced very small quantities of BW. Still, in January 2001, the US DOD listed Syria being among those countries having an active BW program.

Egypt

Egypt was the first Arab country to use chemical weapons, bringing CW to battle in Yemen in 1963. Since that time, the Egyptian government has compiled a substantial CBW capacity in order to counter Israel's military prowess, but also, perhaps, in order to keep pace with other Middle Eastern countries' increased development, production, and stockpiling of the weapons.

Chemical Weapons

In 1962, Egyptian academic research on fluoroacetate,[34] a highly toxic compound that has been variously proposed as a chemical weapon, suggested that Egypt might be developing a chemical weapons arsenal. A year later, President Gamal Abdel Nasser's armed forces used chemicals during their four-year war with Yemen. Although former British chemical weapon stockpiles were available to Egypt, Soviet-made aerial munitions filled with mustard and phosgene were utilized instead. Not wishing to be entirely reliant on foreign munitions, the Abu-Za'abal Company for Chemicals and Insecticides, also known as Military Plant No. 801, was established in 1963 and was clandestinely directed by the Egyptian Ministry of Defense.

In the 1970s, Egypt expanded its range of chemicals to include nerve agents and published research on psychoactive chemicals to be used in war. During the 1980s, Egypt began manufacturing phosphorus trichloride, a key nerve agent precursor, with the help from the Swiss firm Krebs AG.

In addition to acquiring its own chemical weapons, the Egyptian government arranged to give hydrogen fluoride, a key element in the sarin production process, to Iraq via the United Kingdom. Then, in 1981, Egypt sold key CW agent manufacturing technology to Iraq for $12 million. In addition to serving as a supply house for Iraqi chemical weapon production, Egypt gave $6 million to Syria for its CW program in 1972.

In 1992, reports surfaced that Egypt was procuring some 340 tons of chemical weapon precursors from India and Hungary to manufacture nerve agents, and that it had been working with Russian and North Korean engineers to design improved ballistic missiles that could deliver chemical agents.[35] Additional evidence suggests that from raw materials and intermediaries (such as phosphorus pentasulfide, a VX precursor), to the finished product, Egypt almost has a complete indigenous capacity to produce chemical weapons.

While not publicly admitting to possessing chemical weapons, Egypt justifies the need for a CW capability in order to provide a counterweight to Israel's nuclear arsenal. Furthermore, Egypt, unlike Israel, has not signed the CWC and has encouraged other Arab nations to refuse participation so long as Israel retains its nuclear weapons.

Biological Weapons

...The United States believes that Egypt had developed biological warfare agents by 1972. There is no evidence to indicate that Egypt has eliminated this capability and it remains likely that the Egyptian capability to conduct biological warfare continues to exist. . . .

—United States Arms Control and Disarmament Agency, 1996[36]

In the early 1970s, President Anwar Sadat hinted that an Egyptian BW agent stockpile existed, housed in special refrigeration structures, and that Egypt had continued to conduct research in pathogens and toxins of various sorts since at least the 1960s. Suspected BW agents under study or development by Egypt over the past three decades include the diseases and associated pathogens shown in Table 2-1.

Table 2-1. Suspected Egyptian BW Agents

Disease	Pathogen
Anthrax	*Bacillus anthracis*
Botulinum	*Clostridium botulinum* (toxin)
Plague	*Yersinia pestis*
Cholera	*Vibrio cholerae*
Tularemia	*Francisella tularensis*
Glanders	*Burkholderia mallei*
Meliodosis	*Burkholderia pseudomallei*
Brucellosis	*Brucella melitensis*
Psitacosis	*Chlamydia psittaci*
Q-Fever	*Coxiella burnetti*
Japanese B encephalitis	Japanese B encephalitis virus
Eastern equine encephalitis	Eastern equine encephalitis virus
Smallpox	*Variola major*

Egypt's contacts with Iraqi CBW experts in the late 1980s and early 1990s has led one analyst to suspect cooperation in biological weapons research as well. However, Egypt has signed the BTWC, and in 1980 declared that it had never developed or produced biological weapons. At a BTWC Review Conference in 1986, a representative from Egypt stated that, while in basic agreement with the goals of the Convention, Egypt will not ratify the BTWC until a wider regime covering all weapons of mass destruction is created.

Libya

In the past, Libya's zeal for nuclear weapons was a source of international concern. Today, this threat has been greatly diminished, but concerns about Libya's CBW capabilities have emerged.

Chemical Weapons

Libya's first attempt at using poison gas occurred during its brief war with Chad. Like the overall military operation itself, the forays into CW were dismal failures. At one point, following an artillery attack using chemical weapons, Libyan troops were exposed when CW agents were blown back against them.

Security analysts disagree on the source of Libya's nascent chemical weapons technology. During the 1973 Arab-Israeli war, Egypt, West Germany, or both may have contributed to the conflict by giving chemical munitions to Libya. By 1983, West German intelligence had determined that Libya was manufacturing chemical weapons using German technology, but the export of key equipment for the construction of a CW agent facility at Rabta, Pharma 150, continued. The Libyan effort at this facility has not been a significant success: the total amount of chemical agents produced there, about 100 tons since 1992, is relatively minor for a state-level program.

Operation at Tarhunah

Fears of a preemptive strike by the United States against Pharma 150 may have led Libya to build another production site. The Tarhunah facility, built into a granite mountain, was described by then Director of Central Intelligence John Deutch as the "world's largest underground chemical weapons plant."[37] However, between tough talk from the United States, which considered using conventional or even nuclear devices to destroy this facility, and more subtle diplomatic pressure from Egyptian President Hosni Mubarak, Libyan leader Muammar Qaddafi promised to halt construction at Tarhunah. By the summer 1996, US Department of Defense sources confirmed that the site appeared dormant. But the dormancy was fleeting, and in March 1997 Israeli sources reported that construction at Tarhunah had resumed. Not surprisingly, Libya has still refused to sign the CWC.

Satellite reconnaissance provided early circumstantial evidence that Pharma 150 was not a facility producing pharmaceutical precursors or agricultural equipment, as Libya variously claimed. Physical features of the plant, as well as human intelligence gathered elsewhere, all pointed to a CW agent factory. The evidence included high fences with 40-foot embankments, oversized air-filtration systems, corrosion-resistant pipes, and an adjoining metal-working plant capable of making artillery shells.

Additionally, satellite photographs of dead wild dogs, killed by a toxic spill nearby, also indicated that some very noxious agents were being handled at the site. Clinching the overall intelligence picture, an intercepted 1988 phone call made over international telephone lines to Imhausen-Chemie, the German designer, indicated that a highly toxic gas was being synthesized.

Biological Weapons

While Libya may be attempting to develop biological weapons, their efforts are believed to be in the research and development phase, and it is unlikely that they would develop a militarily significant capacity in the near future.

Israel

Although the Israelis are believed to maintain stockpiles of chemical and biological munitions, and Israel certainly has the technical capability to produce them, almost nothing is known in the open literature (unclassified documents) about Israel's CBW arsenal. Analysts have even discounted earlier reports that there was a CW production facility in the Dimona area making nerve and blister agents. However, given its overall military strength, it is highly likely that Israel has both chemical and biological weapons, but at what scale is not known.

Chemical Weapons

Clues about Israel's CW program are, however, gradually being revealed. According to a 1998 report in the *London Sunday Times*, the Israeli Institute for Biological Research is responsible for developing chemical (and biological) warheads, including those for equipage on F16 fighter aircraft on a moment's notice. A 42-gallon shipment of dimethyl methylphosphonate (DMMP)—a nerve agent precursor—was headed for the self-same institute on an El Al airliner in 1992 when the 747-200 aircraft carrying it crashed into an apartment high-rise in Amsterdam. Although DMMP is considered a compound with high proliferation potential, particularly for sarin manufacture, it is used commercially as a flame retardant and stabilizer for plastics. Israeli sources did admit that DMMP was destined for military research, but only as a nerve agent simulant to test protective equipment and detectors. The relatively small amount of DMMP involved suggests the latter explanation is plausible.

Biological Weapons

Very little is known about the nature of Israeli biological weaponry, although they are believed to at least have such weapons. Experts suggest that Israel's BW program may have been modeled on the arsenals of the US and the former Soviet Union. If this is true, Israel probably has stockpiles of anthrax, botulinum toxin, tularemia, plague, Venezuelan equine encephalitis, and Q-Fever, but this level of detail has yet to be confirmed. Like several of its Arab neighbors, Israel has not signed the BTWC.

EAST ASIA

Several countries in this region, most notably North Korea, have developed substantial reserves of both chemical and biological weapons. Any hopes of meaningful international conventions for the control and abolition of CBW must take this region, as vital as any on the globe, into account.

North Korea

According to experts at the Henry L. Stimson Center, North Korea is "one of the most closed and militarized societies on Earth."[38] Its army and defense infrastructure seem to take precedence over any other concerns that might press on the government, even in the face of economic and agricultural crises that have resulted in the starvation of millions of its citizens in the period between 1994 and 1998. Over the years, the North Koreans have developed a large chemical and biological weapons reserve that warrants serious discussion.

Chemical Weapons

... Poison gas is employed to destroy the fighting power of the
enemy army by the deprivation, either temporary or permanent,
of the normal function of a part of the physiological structure of
men or beasts. Again, by making the enemy use protective equipment against poison gas, the combat range of the enemy army is
reduced or curtailed, and their combat strength is weakened. ...

—North Korean People's Army
Chemical Warfare Manual, 1947[39]

North Korea's offensive CW program began in the 1960s, initially relying upon Chinese to help develop the weapons, and upon Japan and other countries to obtain precursors. However, for all of their efforts, the program failed to yield impressive results. Alternately relying on the Soviet Union and China for technical support and aid, North Korea probably produced its first small quantities of mustard and nerve agents in 1966 with the help of the USSR. A serious program of research and development then got underway. However, in the 1970s, the Soviets once again became disenchanted with the country and left it to China to take up the slack. By the 1980s, North Korea was able to produce CW agents in large quantity and deployed large numbers of chemical ordnance.

At one time, North Korea may have hoped that chemical weapons would be an adequate foil to the United States and its nuclear warheads in South Korea. During this period of Kim Il-Sung's adventurism directed at the South, North Korea may have felt that having a CW arsenal made them immune to a forceful response. But after the United States' strong reaction to a North Korean attempt on the life of South Korean President Park Chung-Hee (1968), the seizure of the USS Pueblo (a few days later), and the shooting down of a US Navy EC-121M (April 1969), Kim Il-Sung began to think otherwise, and after a major political purge of his advisors, embarked on biological and nuclear research.

Built upon an ideological platform of communism that emphasizes *juche* (self-reliance), North Korea has threatened to invade Seoul and forcefully unify the Korean peninsula. While on paper the North has numerical superiority in terms of men (and women) under arms, as well as heavily armored divisions, much of its military equipment is in poor shape. As a force-multiplier—at least in the minds of the Northern leadership—chemical weapons may offer a qualitative military edge.

Suspected CW Arsenal

Open-source assessments of North Korean chemical weapons stores have changed markedly over the past decade. Reports by defectors and other accounts in the open literature indicate that North Korea currently possesses the following CW agents:

- Sarin (GB)
- Tabun (GA)
- Soman (GD)
- VX
- VM
- Phosgene
- Mustard
- Cyanogen chloride
- Hydrogen cyanide
- Diphosgene
- Adamsite (diphenylaminochloroarsine, DA)
- BZ
- Diphenylchloroarsine (DA)
- Phosgene oxime (CX)

US Congressional testimony from a recent North Korean defector claimed that "Lizut" (Lewisite), a potent blister agent, was also a part of the North Korean chemical arsenal. The South Korean government has claimed that the North has other V-series analogues, VE and VG, although these claims seem to be based on inference rather than hard data. A 1999 report made to the US Congress by the North Korea Advisory Group said that:

> . . . Reflecting Soviet military doctrine, the DPRK [North Korea] has traditionally viewed chemical weapons as an integral part of any military offensive. There are no indications that this view has altered since the end of the Cold War. The most obvious tactical use of chemical weapons by the DPRK would be to terrorize South Korean civilians. Seoul lies within easy striking distance of North Korea's artillery and rocket systems and, today, the South Korean civilian population has no protection against CW attack. . . .[40]

Because North Korea lacks a certain number of indigenous precursors, they have probably focused on the production of phosgene, mustard, sarin, and V-agents. Previous reports have also suggested the presence of a large nerve agent stockpile consisting mainly of sarin. Aided by an established chemical industry, North Korea has the capacity to manufacture thousands of tons of chemical weapons each year.

However, the recent dramatic downturn in North Korea's economy has led to a severe shortage in both energy and raw materials. Therefore, it is more difficult to estimate production (if any) of chemical weapons in North Korea today. For example, Kim Il-Sung's heir Kim Jung-Il has been forced to accept food and energy assistance from his nemeses, namely Japan, the United States, and South Korea. During the famines and energy shortages of the 1990s, North Korea has moderated its stance in some respects. In 1994, North Korean leaders even agreed to put a temporary halt to its nuclear weapons program.[41]

Biological Weapons

> North Korea has pursued research and development related to biological warfare capabilities for the past 30 years. North Korean resources, including a biotechnical infrastructure, are sufficient to support production of limited quantities of infectious biological warfare agents, toxins, and possibly crude biological weapons. North Korea has a wide variety of means available for military delivery of biological warfare agents.
>
> —*US Secretary of Defense, Proliferation: Threat and Response, 1996*[42]

North Korea has ratified the Biological and Toxin Weapons Convention, but is rumored to still maintain its BW arsenal. What little is known about North Korean chemical and biological weapons has been culled from statements from defectors from North Korea over the years. Although such evidence is fraught with uncertainty, the closed system in North Korea allows for very little reliable information from many other sources. Defector testimony differs as to the extent of its BW agent weaponization, although it is generally believed that the country has had an offensive biological weapon arsenal of some kind since the 1970s.

While North Korea probably received much of its chemical warfare capability from China and the Soviet Union, it is less certain how it obtained BW agents and associated technology. Whatever the origins, an estimated 10 to 13 different types of microorganisms were investigated during the early development process, including anthrax, cholera, plague, smallpox, and yellow fever. Following this initial effort, a laboratory and facility organization were established under the aegis of the Academy of National Defense. However, the results at this stage were lackluster. In 1968, North Korea imported anthrax (*Bacillus anthracis*), plague (*Yersinia pestis*), and cholera (*Vibrio cholerae*) bacteria, presumably obtained from culture collections in Japan. Research into typhoid (i.e., causative bacterium *Salmonella typhi*) has also been reported, although this may have been confused with the causative agent of typhus (*Rickettsia prowazekii*).

The production of BW agents, including the causative bacteria of cholera, typhus (rickettsial), tuberculosis, and anthrax, began in the 1980s. Botulism, the plague, yellow fever, hemmorhagic fevers, and smallpox were researched and developed around the same time. Of these nine biological weapons, experts are more confident that North Korea has produced agents for anthrax, botulism, and the plague. Unlike its CW development, biological weapons development in North Korea has been mostly indigenous, being researched and developed in both civilian and military research institutes.

Even beyond its capacity to produce biological weapons, North Korea's military has a number of deployment options including artillery and ballistic missiles and low-altitude aircraft. The South Korean government estimates that half of North Korea's long-range missiles and 30 percent of its artillery pieces are capable of firing chemical or biological warheads.

South Korea

As a State Party to the CWC, the Republic of Korea was required to fully declare any offensive chemical weaponry, and to the surprise of some, Seoul acknowledged possession of some chemical munitions. The details on what exactly South Korea possesses are still sketchy, but apparently production and weapons facilities slated for chemical ordnance were located near or at the Demilitarized Zone. These weapons could range from VX land mines to possibly older agents that have been geared for a conflict with the North. As a ratifier of the Convention, however, South Korea has committed to the removal and destruction of these chemical weapons and related infrastructure.

China (PRC)

No evidence in the open literature suggests that the People's Liberation Army, or PLA, possesses chemical weapons (beyond those left over from Japan's invasion of China), or that it is prepared to use them offensively. Considering the poor quality of CW defense training, the mismatch of chemical protective gear, and the generally low technical level of the PLA, it is unlikely that the Chinese military could conduct large-scale offensive CW operations. If Chinese writings on chemical weapons are scanty, even less information is available on biological weapons. A book on the subject with the imprimatur of Chinese Defense Minister Chi Haotian states categorically that "China has never manufactured nor possessed biological weapons."[43] Yet according to a 2001 report by the US Department of Defense, "China continues to maintain some elements of an offensive biological warfare program it is believed to have started in the 1950s. . . . China is believed to possess an offensive biological warfare capability based on technology developed prior to its accession to the [BTWC] in 1984."[44]

China has conducted a considerable amount of ostensibly defensive research on potential BW agents, including the causative agents of tularemia, Q-Fever, plague, anthrax, eastern equine encephalitis, and psittacosis, among others. China also possesses the technology to mass-produce most traditional BW agents, including the causative agents of anthrax, tularemia, and botulism. Finally, China has expertise in aerobiology and conducts laboratory-scale aerosolization experiments with microorganisms.[45] Nevertheless, it is not known from open sources whether China has the technology for the efficient delivery of BW agents, or whether field-testing with animals has been performed in the past.

Chemical Weapons
Over time, a number of clues have defined the edges of China's closely-guarded, well-established CW capacity. For instance, China has advanced expertise in chemical agent production, chemical defense measures, and supplies its own armed forces with paper-based and electronic chemical-detection kits. Furthermore, since the 1960s, the Chinese government has

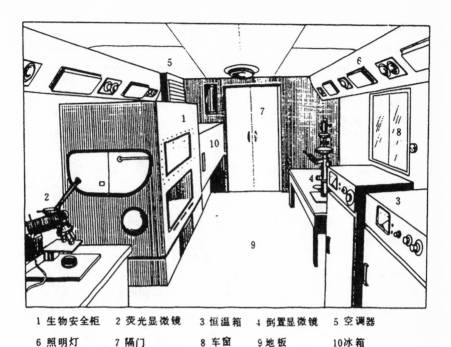

1 生物安全柜   2 荧光显微镜   3 恒温箱   4 倒置显微镜   5 空调器

6 照明灯   7 隔门   8 车窗   9 地板   10冰箱

According to the People's Liberation Army, a mobile BW agent detection laboratory has been fielded by the Chinese military.

1 – Safety/biohazard glove box
2 – Light microscope
3 – Incubator
4 – Inverted light microscope
5 – Air conditioner
6 – Light panel
7 – Compartment door
8 – Window
9 – Floor
10 – Refrigerator

Photograph: From Chinese Military Encyclopedia: CBW Defense
Technologies *(Beijing: PLA Press: 1990).*

emphasized the importance of research on nerve agent antidotes and treatment. In spite of this evidence, the PRC continues to stress the purely defensive nature of their "Anti-chemical corps." Reliant as it is on a sizeable nuclear weapons capability, it may very well be that chemical agents do not form a significant part of China's military arsenal.

Having since signed and ratified the CWC, China is one of eight nations that has "declared possession of existing or former chemical weapons production facilities."[46] In 1998, following allegations that the PRC had played a role in the proliferation of CW technologies, China voluntarily added ten chemical precursors from the Australia Group list to the list of chemicals it will control for export. But even if China itself is not heavily invested in a CW program, they appear to be helping other nations establish their own chemical weapons arsenals. In 1993, the US government alleged that a Chinese registered vessel, the *Yin He*, was transporting mustard and nerve agent precursors to Iran. After a lengthy standoff, inspectors failed to find any of the offending chemicals, although some suspect that, as soon as Chinese officials got wind of the allegations, the precursors were surreptitiously unloaded at Jakarta. In 1997, Hong Kong-based companies shipped 500 tons of phosphorus pentasulfide to Iran, a well-known precursor to VX as well as an intermediate for pesticides. Again, the Chinese government claimed that an export of dual-use chemicals was not possible because of its efficient export controls. Later, however, it claimed that there are more than 6000 chemical enterprises in China, which made it impossible to track commercial activity in this sector.

Biological Weapons

According to the US Department of Defense, China "may well have maintained the biological warfare program it had prior to acceding to the Biological Weapons Convention in 1984," a program that "included manufacturing infectious micro-organisms and toxins."[47] A more harshly written estimate of China's BW capability was released in 1995, when—during a diplomatic row over Taiwanese President Lee Teng-Hui's visit to Cornell University in upstate New York—the United States claimed that the Chinese military had continued to research defensive and offensive uses of BW agents. So far, the Chinese government denies running any offensive program in BW, asserting that it only pursues defense-related research work and investigations of infectious disease of public health concern.

Taiwan

Although its government officials continue to deny the charge, Taiwan is often listed as a country with offensive CBW capabilities. It is possible that some remnants of weapons left after the Japanese occupation during World War II may have been utilized by the Nationalist government after the 1949 Chinese revolution. No details are available on which agents (if any) Taiwan has weaponized. Chemical and biological defense is, of course, a topic of research in Taiwan, although here too the extent is unknown.

OTHER PLAYERS

In addition to the countries previously discussed, there are two others whose chemical and biological weaponry merit discussion here—South Africa and Cuba. Hard information on either is difficult to come by, and neither is currently a major player in CBW. Nonetheless, both played a significant role in the past, and both may, to this day, be harboring CBW armaments.

South Africa

South Africa's apartheid-era chemical and biological weapons program flourished under the leadership of Wouter Basson, who is currently on trial before the Truth and Reconcilliation Commission for atrocities committed in the course of his work on CBW.

Chemical Weapons
In November 1963, South African scientists began to investigate and plan the development of CBW technologies with the help of West German industry. The emphasis at first was on riot agents. In the mid-1980s, the South African Defense Forces (SADF), a pro-apartheid movement, used front companies, including Roodeplaat Research Laboratories and Delta G Scientific, to develop South Africa's chemical weapons program. Basson, a former military surgeon for the SADF, was arrested for selling illegal Ecstasy[48] tablets in the beginning of 1997. When it was learned that Basson was also the founder and director of South African CBW programs—code-named Project B (later Project Iota) and Project Coast—another investigation was launched by the Truth and Reconciliation Commission, probing SADF's chemical weapons activities.

Testimony by former members of Basson's team of scientists revealed the chilling nature of South Africa's CBW program—an effort that was largely aimed at killing political dissidents within its own borders. For example, the South African military planned to poison Nelson Mandela with thallium, an extremely toxic heavy metal. In order to suppress political opposition, poison-embedded T-shirts were to be distributed among anti-Apartheid activists. According to the plan, a toxin in the shirts would penetrate the skin, causing death by triggering blood clots and cardiac failure. Fortunately, the police squads charged with carrying out with this operation lost their nerve, abandoning the operation at the last minute. Basson's unit also produced drugs such as mandrax (methaqualone, a barbiturate) as well as Ecstasy ostensibly for riot control and interrogation, but more likely for the illicit drug trade.

Because South Africa's Truth and Reconciliation Commission (TRC) promised the South African Defense Minister not to divulge information that could compromise national security, more detailed information about South African chemical weapons has not yet been made available.

Biological Weapons

Testimony before the TRC also revealed that the apartheid-era South African government used biological weapons to murder political opponents and to "maintain peace and stability."[49] While many of the activities of the so-called Project Coast headed by Wouter Basson remain secret, witnesses told the Commission about producing umbrellas with poisoned tips and making anthrax-infected chocolates and cigarettes. A leading scientist admitted that he tried to develop a serum that would render black women infertile. At the government's request, other scientists tried to create a bacteria that would only kill or injure black people. Thankfully, their efforts failed.

Cuba

If data are sparse concerning CBW capabilities in a number of "countries of concern," this is even more so in the case of Cuba. A protectorate of the former Soviet Union, Cuba participated in regional conflicts such as the Angolan campaigns in the 1980s, when charges were made that it possessed CW capabilites. None of these charges seems to have much foundation. If there is an area in which Cuba is more likely to have some expertise, it is in biological weapons. But again, no open literature has been able to confirm or deny Cuban involvement in offensive CBW research.

SUB-STATE ACTORS

In April 2001, the US Department of State released a document entitled "Patterns of Global Terrorism—2000," published by the Office of the Coordinator for Counterterrorism, and reflecting the official views of the US Secretary of State. In the report, seven nations—Iran, Iraq, Syria, Libya, Cuba, North Korea, and Sudan—are designated as "state sponsors of international terrorism," with each given a kind of thumbnail portrait:

... Iran remained the most active ... increasing support to
numerous terrorist groups, including the Lebanese Hizballah,
HAMAS, and the Palestine Islamic Jihad (PIJ). ... Iraq continued
to provide safehaven and support to a variety of Palestinian rejec-
tionist groups, as well as bases, weapons, and protection to the
Mujahedin-e-Khalq (MEK), an Iranian terrorist group that
opposes the current Iranian regime. Syria continued to provide
safehaven and support to several terrorist groups. ... Libya at the
end of 2000 was attempting to mend its international image following its surrender in 1999 of two Libyan suspects for trial in
the Pan Am 103 bombing. ... Cuba continued to provide safehaven to several terrorists and US fugitives. ... North Korea har-
bored several hijackers of a Japanese Airlines flight to North
Korea in the 1970s. ... Finally, Sudan continued to serve as a
safehaven for members of al-Qaida, the Lebanese Hizballah, al-
Gama'a al-Islamiyya, Egyptian Islamic Jihad, the PIJ, and HAMAS,
but it has been engaged in a counterterrorism dialogue with the
United States since mid-2000. ... [50]

This list, updated and issued every two years by the US State Department,
is intended as a mechanism for publicly identifying and isolating state sponsors
of terrorism. It explicitly declares that it is aimed at making designated states
not just renounce terrorism and end material support to terrorist organizations, but bring individual terrorists—whether their own citizens or
"guests"—to justice for past crimes. The report goes on: "The United States is
committed to holding terrorists and those who harbor them accountable for
past attacks, regardless of when the acts occurred. The US Government has a
long memory and will not simply expunge a terrorist's record because time
has passed." The document makes for sobering reading, and is striking not
only in its forceful and determined language about terrorist adversaries, but in
a certain business-like acknowledgment of the progress being made by some
states known to have supported terrorism in the past. Even Libya, for example,
is credited with having handed over two terrorists who were convicted of
murder and conspiracy for the bombing of Pan Am Flight 103, and North
Korea and Sudan are mentioned as having gone some way toward renouncing
their connections to terrorist organizations.

What is also striking about the statement, however, is how far it is from
providing a clear way out of our present dilemma. Suppose we accept the fact,
now apparently undeniable, that it was Al Qaeda, Osama bin Laden's organization, that was the driving force behind the September 11, 2001, attacks on the
Pentagon and World Trade Center. Suppose also that we believe Al Qaeda is
likely to organize, or at least attempt to organize, a CBW attack on the United
States or on its citizens abroad. If we then compare the US State Department's

list of state sponsors to our own list of likely CBW proliferators, we see that there is no correspondence. The State Department's list points to Sudan as a likely sponsor of Al Qaeda ("al-Qaida"), but Sudan is not known to possess either chemical or biological weapons capacity. And if one relies instead on newspaper and television reports, and agrees with what seems obvious—that it is Afghanistan that is the sponsor (or at least the host) of Al Qaeda—the connection seems even more problematical and far-fetched. Not only is Afghanistan not on the State Department's list, but it is clearly incapable of providing CBW armaments, and in fact, after the tragic turmoil of the last 20 years, is barely able to operate as a state. Where, then, is Al Qaeda likely to get its chemical or biological weapons, if it is to get them at all?

What turns out to be the case is that Al Qaeda, unlike many earlier terrorist organizations, is not a tightly organized group. It is not a clearly defined organization with a single sponsor. Instead, it is a loose network, and serves as more of a clearinghouse and fundraiser and coordinator of widely spread and compartmentalized cells of operatives, many of whom may even be unaware of each other's existence. Rather than looking directly to a single sponsor state for logistical and financial support, Al Qaeda works in the shadows, securing funds from (or at least through) wealthy and sympathetic individuals and non-governmental organizations—charities, foundations, and commercial organizations. We call groups like Al Qaeda "sub-state" organizations, but it is also important to understand that they are trans-national. With members from many different countries, and residing in many different countries, such a decentralized and loosely bound organization is extremely hard to define, to penetrate, and to attack. Our question here, though, is this: Can such a shadowy organization mount a credible CBW attack?

In a prescient report written at the request of Congress by the National Commission on Terrorism and published in June 2000, the authors not only predicted that there would, before too long, be a major attack on US soil, but that the very nature of terrorist organizations had changed significantly in the last twenty years:

> . . . Most terrorist organizations active in the 1970s and 1980s had
> clear political objectives. They tried to calibrate their attacks to
> produce just enough bloodshed to get attention for their cause,
> but not so much as to alienate public support. Groups like the
> Irish Republican Army and the Palestine Liberation Organization
> often sought specific political concessions. . . . Now, a growing
> percentage of terrorist attacks are designed to kill as many people
> as possible. In the 1990s a terrorist incident was almost 20 percent
> more likely to result in death or injury than an incident two
> decades ago. The World Trade Center bombing in New York

killed six and wounded about 1000, but the terrorists' goal was to topple the twin towers, killing tens of thousands of people. . . .

—from "Countering the Changing Threat of
International Terrorism"[51]

If it is the case that sub-state actors are turning to CBW agents, and their goals have shifted toward more and more casualties, we will have to rethink how we will limit proliferation. Because the world in which they operate is clandestine, and immune to the sanctions we have in place for states, we probably cannot put much stock in the current CBW disarmament regimes.

CHAPTER 3

Threats and Responses

We learn from history, but making predictions from history is another matter altogether. Almost exactly two months before the September 11, 2001, attack on the Pentagon and World Trade Center, a knowledgeable and deservedly respected expert on terrorism wrote a *New York Times* Op-Ed piece, "The Declining Terrorist Threat," in which he said the following:

> . . . Americans are bedeviled by fantasies about terrorism. They
> seem to believe that terrorism is the greatest threat to the United
> States and that it is becoming more widespread and lethal. They
> are likely to think that the United States is the most popular target
> of terrorists. And they almost certainly have the impression that
> extremist Islamic groups cause most terrorism. None of these
> beliefs are based in fact. . . .[1]

As perilous as prediction is, in this chapter we try to arrive at a reasonable estimate of just how serious the threat of CBW terrorism is. We can't of course say with any degree of accuracy that there will be this or that number of attacks, but we do try to arrive at a realistic sense of how effective—how widespread and lethal—such attacks are likely to be.

A NEW KIND OF WARFARE

Over the last few years, military theorists have spent a good deal of time thinking and talking about "asymmetrical" warfare, roughly defined as warfare between forces that are very different from each other in size, in technological sophistication, even in goals. The notion of two similarly equipped and trained armies, facing each other over a territory each wants to occupy, is replaced by adversaries who bear little resemblance to each other—a small mobile force against a large standing army, a lightly armed guerilla operation against a heavily mechanized professional force. Terrorists are, in a a sense, the most daunting asymmetrical opponent for a typical modern army. Not only are their tactics and weapons completely different from those of a conventional military force; the terrorists don't necessarily even engage in battle with the army, instead taking their attacks directly to civilian targets.

The First World Trade Center Bombing

In 1993, terrorists bombed the World Trade Center in New York City. Though the extent of physical damage was great and six people lost their lives, the results were far short of what the bombers hoped for. Before long, in the popular press in New York, the terrorists were mocked for their incompetence. The plotters had hardly made a dent, it seemed (even though six people died), and they were incredibly inept at covering their trails. The truth about the attack, however, turned out to be more chilling than laughable. It was later discovered that the leader, Ramsi Yousef, was planning on causing 250,000 casualties by making one tower fall into the other, and his operatives, for all their failings, very nearly succeeded. Moreover, Yousef was far from incompetent, and unlike his co-conspirators, he was not caught—at least not immediately.

As we have learned subsequently, Yousef considered using a chemical agent in the bomb. On a flight back the United States two years later, following his capture in Pakistan by US authorities, he told a Secret Service agent the original plan was to include cyanide in the explosion. What stopped him was the fact that devising the means to deliver the poison "was going to be too expensive."[2] This was not, apparently, an idle boast. A few months after the bombing, Yousef said this in a letter describing another operation:

> . . . We [have] the ability to make and use chemicals and poisonous gas. And these gases and poisons are made from the simplest ingredients, which are available in the pharmacies; and we could, as well, smuggle them from one country to another as needed. And this is for use against vital institutions and residential populations and drinking water sources and others. . . .[3]

The chemicals and components in the Trade Center bomb cost less than $15,000.[4] How much more money Yousef would have needed to add cyanide to the bomb, or how he planned to engineer such a device, is not known. But certainly CW armaments were on this terrorist's agenda.

Aum Shinrikyo

Probably the most widely reported instance of CW terrorism was the 1995 assault on the Tokyo subway by Aum Shinrikyo, a global apocalyptic religious cult based in Japan. During the morning rush hour on March 20, 1995, Aum operatives boarded five separate subway trains carrying plastic bags containing the deadly nerve agent sarin. At a predetermined time, they punctured the bags with umbrella tips, and the poisonous agent was released.

Previous attempts by Aum to gas victims were frantic and almost comical failures, except for the fact that they killed people. The cult had a history of using nerve agents in terrorist operations. In a particularly egregious 1994 attack, they rigged a refrigerated truck with an apparatus to spray sarin on several magistrates who were on the verge of ruling against them in a legal dispute. The judges, though injured, managed to survive, but 7 bystanders were killed, and 144 were seriously hurt.[5] In the year that followed, because of heavy police scrutiny, Aum had a hard time maintaining the size and potency of its stockpiles of sarin, and by the time the cult decided to make its assault on the Tokyo subway, supplies had run so low that it needed to dilute the agent with the solvent acetonitrile. This increased both the volume and the volatility of the liquid in each bag, but it also diluted the sarin. When the bags were punctured, the mixture seeped up and out, then evaporated. This method of delivery was certainly crude, but 12 people were killed, and approximately 1000 injured in 16 stations throughout the subway system. This was a horrible outcome, but it could have been much worse—by mixing the agent with the solvent, and with sarin that was of poor quality to begin with, Aum had greatly reduced the toxicity of their weapon.[6]

The Aum Shinrikyo case is instructive. It demonstrated that even without state support, independent actors can acquire large amounts of chemical precursors. It has been estimated that at one time Aum possessed 50 tons of phosphorus trichloride, several 55-gallon drums of isopropyl alcohol, and some 10 tons of sodium fluoride. These ingredients, with some other precursors easily purchased in the chemical industry, could have been used to make hundreds if not thousands of kilograms of agent—almost certainly ten times what the cult had at its disposal at the time of the subway attack.[7]

While the chemicals involved in synthesizing nerve agents are harder to acquire than Ramsi Yousef would have one believe ("the simplest ingredients . . . available in the pharmacies"), they are nonetheless available, particularly if one is willing to pay a premium and work through illicit supply channels. And with a few decent scientists (Aum at one time had several working in its labs), these precursors can be put into weapons.

Taking the Toll

What's one to make of these two incidents? Is it only a matter of time before terrorists like Aum and Yousef master the technical details of chemical and biological warfare? Is it inevitable that they will eventually manage to kill not dozens, but thousands of civilians in a mass attack with biological or chemical agents? Or is there something holding them back?

As terrifying as the prospect of chemical or biological attacks may be, one needs to keep in mind that the number of people killed by CBW terrorists is minuscule compared to the number killed by terrorists wielding bombs. There is no real mystery why. First, explosives have already amply demonstrated their destructive power in the history of warfare. They are well known, reliable, and predictable. They are relatively easy to manufacture and deploy. They mostly remain stable during manufacture and transport. And they are, in a sense, "obedient"—you place a bomb where you want it to explode, and you detonate it.

Familiarity, safety, predictability—these are the attractions of conventional armaments. The 6500-pound bomb built by Timothy McVeigh and Terry Nichols, which created such devastation in Oklahoma City in 1995, was relatively easy to make, largely from ingredients that were perfectly legal to buy. McVeigh purchased 4000 pounds of ammonium nitrate fertilizer (the oxidizing component) and three 55-gallon drums of nitromethane (the fuel component). The only "controlled" component he needed—the sensitizing charges—he stole from a rock quarry storage facility.[8] The rest was horribly simple. He loaded the bomb into a truck, drove it to the Alfred P. Murrah building, parked directly in front, then lit the fuse and walked away. The blast destroyed the entire structure and killed 168 people—among them children in a day-care center. The explosive could hardly have been more low-tech.

Conventional munitions were also the weapon of choice in 1998, when Al Qaeda operatives set off truck bombs at two US embassies, in Tanzania and Kenya. Though much smaller in size than the Oklahoma device, these bombs employed higher-energy explosives (TNT). The explosion at Dar Es Salaam killed 11 and wounded 72. The bomb in Nairobi killed 212 and wounded more than 4600. And the weapons in the September 11 attack, although by no means "conventional" in terms of delivery, were diabolically simple: aircraft fully loaded for transcontinental flight with fuel that exploded on impact.

In our fear of novel and poorly understood CBW weapons, it's important not to lose sight of the fact that explosives will, at least in the foreseeable future, remain the terrorist's weapon of choice. While it's absolutely true we have to think about clouds of anthrax germs or mists of sarin nerve agent, it's just as urgent that we pay attention to that large truck parked in a suspicious spot, or even more disturbingly, the report of yet another unstable regime brandishing its newest nuclear warhead.

MEDICAL THREATS AND RESPONSES

To many, the most insidious form of terrorism is the one that threatens us with disease. Fear of biological agents in particular makes people nervous, and it is not hard to see why. Biological agents may make us sick, even sick unto death, before we even know we've been attacked. They are also particularly disturbing because they take one of the greatest and hardest-won achievements of modern science—its understanding of how to overcome disease—and turns it completely around. Diseases that were eradicated—wiped off the face of the earth—are brought back to haunt us. Diseases that have had no significant effect on the health of the public for decades return with a vengeance, more virulent than they ever were in nature.

The consensus among experts in bioterrorism is that there are two agents most likely to be used by terrorists. The first is *Bacillus anthracis*, the bacterium that causes anthrax. The second is *Variola major*, the virus that causes smallpox. In this section, we'll take a brief look at just these two agents. (Chapter 7 has more technical details on these and other biological agents, such as botulism, plague, and tularemia.) But here we'll concentrate on how anthrax and smallpox measure up as threats, and examine the steps that have been taken to respond to their possible use in attacks.

Anthrax

From the perspective of the BW armament designer, anthrax has a number of advantages. First, it has the ability to survive a wide range of environmental stresses in its form as a spore. (A spore is a "dormant" bacterium that has grown a thick wall around itself, for protection from extremes of temperature and other environmental conditions.) Anthrax spores release bacteria upon reaching a good growth environment, such as the nutrient-rich body of a living human. Spores are difficult to kill, and therefore as weapons they have the very desirable quality of remaining stable for long periods. (In fact, anthrax spores that have lain dormant for decades can be brought back to life.)

The Disease
The disease caused by the anthrax bacterium actually comes in three forms—cutaneous, pulmonary, and gastrointestinal (on the skin, in the lungs, and in the digestive tract). The causative agent in all three—*Bacillus anthracis*—remains the same, but its route into the body is different with each form of the disease. By far the most common form of anthrax disease is cutaneous (at about 95 percent of all cases)—often found in people who work with animals like cows, goats, and sheep. Of those who get infected with cutaneous anthrax (usually through abrasions or cuts) 10 to 20 percent are likely to die if the dis-

ease is left untreated. For pulmonary, or inhalation, anthrax, the rate is much higher. This form of the disease occurs when an individual breathes in thousands of spores, and develops an infection that begins in the lungs. Left untreated, some 90 to 100 percent of people who contract inhalation anthrax may die. The gastrointestinal form of the disease, while quite serious (with death rates of roughly 50 percent) is even more rare. It is contracted by eating tainted meat.

From 1900 to 1978, only 18 cases of inhalation anthrax were reported in the United States, two of these being laboratory-acquired infections. But if BW armaments of any consequence are ever used against large populations, it is the inhalation form of the disease that is most likely to be deployed.

From an infected person's point of view, there is some good and some bad news about anthrax. The good news is that it is not contagious (it is not transmitted from person to person) and that it can be successfully treated with antibiotics. The bad news, especially for those who are infected with the inhalation form of the disease, is that they may not know they are infected until they are well into the course of the infection. This is a very serious problem because anthrax needs to be treated quickly. There is a good chance that those infected by inhalation in an attack will interpret their symptoms—cough, chest pains, aches and fever and fatigue—as nothing more than a bad cold or case of the flu. Such a misreading could have terrible consequences, since a delay of even a day in starting treatment may significantly reduce an infected person's chances of survival.

Anthrax as a Weapon

Many statements about the potential devastation of a massive anthrax attack are misleading and unnecessarily alarming. It is indeed true that a mere 2 grams of dried anthrax spores, distributed evenly as a powder over a population of 500,000 in an urban setting, could cause death or serious illness for 200,000 people. Before this happened, however, the attackers would have to overcome a number of extremely challenging technical hurdles:

- First, they would have to have access to a strain of the bacterium that is particularly virulent, even perhaps engineered to resist many antibiotics. Spores harvested from the ground or from farm animals will not meet these criteria; instead, the attackers would have to get hold of a strain that had been deliberately cultivated.
- Next, the attackers would have to "mill" or otherwise "grind down" the supply of spores so that individual particles were small enough to be deeply inhaled, but not so small that they would be immediately expelled in the normal cycle of breathing.

• In addition, the attackers would have to add some kind of anti-static component to the milled particles, because in the process of milling they tend to become electrostatically charged, and thus form into large clumps of spores that would otherwise drop harmlessly to the ground.

Only then, with a stock of "fluffy and fine" anthrax, would the attackers be ready. And then the challenge would be to deliver the material so that it formed an aerosol cover that was large but also sufficiently concentrated, and doing this on any kind of scale requires near-perfect weather conditions—that is to say, with little wind, and few up- or down-drafts to move the aerosol in unexpected directions, or to disperse it. None of these hurdles are insurmountable, but they are hurdles, and the smaller and less resource-rich an attacker, the less likely he is to be able to undertake an operation of this kind on a massive scale. Unfortunately, as we have seen in recent incidents, attacks on a smaller scale, in what now appear to be indoor rather than outdoor settings, have been lethal, and are of course deeply disturbing to citizens everywhere. But the point remains: a terrorist attack on a city, with an immense deadly aerosol cloud descending on thousands of inhabitants, does not seem a likely scenario.

Treatment and Vaccination

The antibiotics ciprofloxacin ("Cipro"), doxycycline, penicillin, and perhaps others are effective against anthrax if given soon enough—meaning by most estimates within just days of exposure. As time passes, however, the likelihood that antibiotics can offer any help diminishes rapidly.

What about the vaccine for anthrax? Despite claims to the contrary, all available evidence shows that the FDA-licensed anthrax vaccine used for US military personnel is safe and effective. At present, the US military is partly through its program to have all of its 2.4 million soldiers and reservists vaccinated. Recently, some have suggested widening the vaccination program to include the general public. Here, too, one needs to consider the risks versus benefits, particularly when a mass release of anthrax spores on a civilian population is still a relatively remote possibility. The vaccine is not without its adverse reactions, and as currently configured requires six inoculations and annual boosters. It is certainly not "user-friendly," and undertaking a program to conduct a blanket vaccination of every man, woman, and child in the country would be fraught with problems. Even if the anthrax vaccine were made available to the public (which as of this writing is not the case), individuals would want to think carefully before going through the time and effort. A nation-wide anthrax vaccination program, albeit unlikely to ever be implemented, deserves a frank and public discussion before it is put in place.

Smallpox

The death rate from inhalation anthrax is close to 100 percent for those left untreated. For smallpox the death rate is considerably lower—perhaps 30 to 40 percent—but no effective treatment currently exists, and worse still, the disease is exceedingly contagious. Smallpox has ravaged human society at least from the beginning of recorded history, killing untold hundreds of millions and scarring for life untold hundreds of millions more. If ever there was a disease that deserved the name "scourge," smallpox is it.

It is all the more remarkable, then, that by 1980 the World Health Organization (WHO) was able to announce a stunning event in the history of human medicine: Scientists, doctors, and public health workers, after decades of concerted global effort, had used vaccines to effectively eradicate smallpox from the planet. It was a stunning achievement. The scourge was no more.

The Disease

Smallpox can be contracted through breathing or by contact with the skin of an infected person. After an incubation period lasting approximately two weeks, sufferers come down with fever and aches. A few days later, blisters appear, and fill with pus, and then burst, causing painful itching. Those who survive the disease are left with extensive pockmarks. While some experimental treatments are under investigation, and while some have suggested that vaccination very shortly after exposure may prevent the disease, there is currently no known proven treatment. And the problem is compounded: smallpox is now potentially more devastating than at any previous time in history. Because routine vaccination in the United States ceased in 1972, and because vaccinations given before that time are by now ineffective, virtually all the world's civilians are susceptible to the disease. Particularly considering the mobility of the global population, along with the fact that in its first two weeks of incubation the disease shows no symptoms, a widespread surreptitious outbreak could have devastating consequences worldwide.

Smallpox as a Weapon

Today, smallpox virus is known to exist for certain in two places: in stores maintained by the US Centers for Disease Control and Prevention (CDC) in Atlanta, Georgia; and at Vector, a biological research laboratory in Novosibirsk, Russia. The original WHO plan called for both the United States and Russia to destroy these stocks of virus, but the plan, from its inception, was controversial. Fears that other nations may have secret stockpiles of the disease prompted WHO to postpone this action, until at least 2002, since samples of smallpox might be needed for research in the event of an attack or outbreak. Given the current concern with biological warfare, and in particular with the possibility that both Iraq and North Korea and perhaps others have stockpiles, it is unlikely that a firm date for destruction will soon be set.[9]

Vaccination

For all intents and purposes, the last case of smallpox in the Unites States appeared in the early 1960s, and routine vaccinations of the general public were ended in the early 1970s. Among US military personnel, however, vaccination programs continued for almost another 20 years. This extension was partly due to the fact that so many personnel were posted overseas, but it was also done as a precaution in case of a Soviet attack with smallpox pathogens.[10]

The subject of vaccination, as always, became a controversial issue. In 1986, some authors suggested that both the former Soviet Union and the United States negotiate an end to their respective smallpox immunization programs. Ending the vaccination requirement would, according to these authors, have had "reassuring implications for reduced biological warfare risk" and "would be a final step in ending the fear of smallpox."[11] Advocates of these measures based their argument, however, on an incomplete knowledge of the prevalence of smallpox throughout the world. First, they assumed only the United States and the Soviet Union possessed the virus. And second, they did not know (as almost no one knew or even imagined), that the Soviets had weaponized smallpox.

Perhaps as early as the 1970s, the Soviet Union possessed at any given time some 20 tons of smallpox virus ready to be deployed in weapons. Russian military scientists had stockpiled huge quantities of smallpox to be loaded onto intercontinental ballistic missiles.[12] In retrospect, knowing as we do now that the Soviets were prepared to use smallpox as a weapon, it certainly made sense that US military personnel were immunized. But in the meantime, the civilian population in the United States, having been originally targeted by Soviet strategic rocket forces in the first place,[13] would have been most vulnerable in a concerted attack.

To prepare against the threat of smallpox used as a weapon, in 1999 the US Working Group on Civilian Biodefense recommended 40 million doses of a new and safer vaccine be produced.[14] More recently, a government proposal was issued in January 2000 to seek out industry participants for such a venture.[15] In September 2000, the CDC contracted Peptide Therapeutics (based in the United Kingdom) to produce this lot of vaccine. And since the increased threat of BW terrorism in late 2001, the US government has increased its order for the newly formulated vaccine from 40 million to 300 million doses. Moreover, current stocks of the traditional vaccine—*Vaccinia*—stand at about 15 million doses, and in the short term these could be used, in diluted form, to provide protection to some 100 million individuals.

Many people have asked, should we all have smallpox vaccinations, or at least be given the opportunity to be vaccinated? As with the anthrax vaccine, this is a complicated issue. When the risk of contracting smallpox in the 1950s and 1960s was high, mandatory vaccination for everyone was a reasonable measure, especially for the goal of eradicating a disease that claimed 500 million lives in the twentieth century alone. Since the disease was eradicated, however, it made no sense to vaccinate everyone. For the time being, at least until the new safer vaccine is available, it probably makes sense to hold off on a mass campaign. But of course in the event of a smallpox outbreak or attack, the risks will change dramatically, and then the benefits of broad-based vaccination to contain an epidemic will have to be reconsidered.

Responses to BW Attacks

The key to saving many lives after a BW attack is early detection, which most likely will occur in an emergency room, physician's office, or health clinic. The so-called *pathognomonic,* or distinctly characteristic signs of a disease like anthrax or smallpox, may today be a bit obscure for health practitioners, but that situation is not likely to last for long. Leading the effort to educate health-care professionals about the tell-tale signs of BW diseases is the Centers for Disease Control and Prevention—the CDC. This federal agency, part of the US Department of Health and Human Services, is charged with and already providing state and local health departments with information on how to detect, diagnose, and respond to potential BW attacks. The CDC also backs up local and state health departments with training, laboratory work, and diagnosis and etiology reports, and publishes the *Morbidity and Mortality Weekly Report* (MMWR), a kind of scorecard that presents and analyzes current trends in the relative incidence of a particular disease (morbidity) as well as fatalities (mortality) and their causes.

A key part of the CDC's mission is communication, so that those on the front lines of the public health system can stay alert to the latest epidemiological trends. This may mean, in the short run, that a kind of high index of suspicion is necessary. For example, in less trying times, if a patient complained of flu-like symptoms to his or her physician, and the physician then said, "This sure looks like the influenza bug that's been going around, but I'm going to run a diagnostic for smallpox just in case," the doctor would be out of line. Physicians are not supposed to look for the least likely cause of illnesses. At the same time, if intelligence sources indicate that a smallpox attack was imminent, or if other cases have occurred elsewhere, it would make sense to keep smallpox or whatever else is currently under suspicion as a so-called "differential" diagnosis, and the CDC would take the lead in making this known.

The CDC has also established the National Pharmaceutical Stockpile (NPS). Located at eight locations throughout the United States (unpublished for security reasons), the NPS contains more than 400 tons (in 50-ton allotments) of antibiotics, including ciprofloxacin but also streptomycin and gentamycin for Gram-negative threats like plague and tularemia. Also included in these stockpiles are nerve agent antidotes, respirators, vaccines, and supplies for fluid administration. The goal of the current NPS as currently configured is to have such emergency medical supplies in "Push Packages" no more than 12 hours from any major populated area in the United States.

Should a terrorist attack involve chemical agents, such as the highly toxic organophosphate compounds sarin, VX, or tabun, the NPS stockpiles also could be used to aid immediate casualties with atropine injections (which can be used to counteract the effects of some nerve agents), respirators, and other treatments. In addition, hospitals and clinics are routinely well supplied with these and other treatments that can save many lives very quickly. In the case of VX, however, risk of contamination to so-called first responders will be great, and decontamination measures must be considered lest more people become inadvertently exposed.

The CDC has also worked with other agencies to develop training courses to prepare medical and emergency teams for what they might face in a CW or BW attack. Those who have gone through simulation training have a better sense of who is in charge (or who isn't, in some cases), who to call, and where the emergency vehicles should be directed. Basic information to be sure, but essential. By going through the "war-gaming" exercises, some mundane but very important data can be gleaned that can save lives in the future.

One important detail of early response has to do with the distribution of antibiotics in the event of a massive biological attack. While enough antibiotics may be available for the affected population, how would they be distributed to individuals? Would they drive to points of distribution or would the drugs be delivered on their doorstep? How could hoarding be prevented? All of these factors must be considered before the public at large can feel confident that the United States is truly prepared.

Another key agency that is already playing a role in the response to BW attacks is not a civilian but a military organization. The US Army Research Institute of Infectious Disease, or USAMRIID, based at Fort Detrick, Maryland, has already contributed significant expertise to the analysis of current anthrax samples. The Institute is part of the US Army, and thus is under the umbrella of the Department of Defense (DOD). As the lead laboratory for investigating biological warfare issues since the US abandoned its offensive BW efforts in the 1970s, USAMRIID has conducted extensive studies of defensive measures, including the development of vaccines, drugs, and diagnostics. The main goal of the scientists at Fort Detrick is, of course, to protect members of the US armed forces, but it has made significant contributions to civilian health efforts, and collaborates on a regular basis with the CDC, in Atlanta, as well as WHO, in Geneva, and universities and research laboratories worldwide. The role that USAMRIID played helping to solve a civilian public-health mystery, the outbreak of West Nile virus in New York City in the fall of 1999, is an object lesson on how a defense agency can contribute to a civilian effort. In this case, crows and other birds, including many rare specimens in the Bronx Zoo, became ill with a mosquito-borne virus. Confusing the issue was the fact that the virus was similar in many ways to St. Louis encephalitis virus, and as a consequence it took some time before West Nile was conclusively implicated in the deaths of New Yorkers. Fortunately, a determined veterinary pathologist at the Bronx Zoo and a diagnostic test developed by scientists at USAMRIID helped solve the riddle before more lives were lost.

USAMRIID has also joined efforts to test and develop advanced BW "alert" technologies in conjunction with DARPA, the DOD's Defense Advanced Research and Planning Agency. (This is the agency that played a key role in conceptualizing and developing early versions of the Internet.) DOD scientists have created prototype laser detection and ranging (LIDAR) devices and other sophisticated detection and diagnostic equipment, but so far no practical system that can quickly confirm or deny that a particular kind of BW attack is in progress. Among many other challenges is the difficulty of designing a machine that can almost instantly distinguish between normal biological "background clutter" (molds and pollen, for example) and particles that would indicate the onset of an actual attack with BW agents. The reality is that civilians, for the foreseeable future anyway, will probably not have access to sensors that can give advance warning of a bioterrorist attack.

But again, it is important to recognize that we are far from undefended against BW attacks. The highly complex and dynamic public health system in the United States, with its complex interrelationships of federal, state, and local agencies (and including a multitude of government, corporate, and volunteer organizations) is extremely adaptable and highly trained. To be adequately prepared will of course require fiscal resources. In 2001, the United States allocated over $340 million for BW defense, more than half devoted to related civilian defense measures. In terms of defending civilian populations from a bioterrorist attack, adding monies to the public health sector seems to have found nearly universal approval. As it stood in 2000, public health received about $250 million, compared to the $3 billion overall spent for measures against chemical, biological, and computer infrastructure attacks. Now is almost certainly the time to spend more. But even admitting that additional funds and coordination are needed in our multifaceted and complex public health infrastructure, it is important to acknowledge the ingenuity, resourcefulness, and determination of those who work on its front lines.

CIVIL DEFENSE THREATS AND RESPONSES

In modern industrial and post-industrial societies, the quality of life—and indeed life itself—is dependent on highly complex and interrelated systems of power transmission, transportation, public and private buildings, commerce, and communications. Terrorists are capable of attacking any or all of these systems. In this section, we'll briefly survey just four potential threats—to the chemical industry, public buildings, water supplies, and agriculture—and assess current threats as well as some in-place and forthcoming responses.

The Chemical Industry

The US federal Clean Air Act of 1990 required, among other things, that information concerning the potential of an accidental chemical release be made available to the public. Thousands of chemical producers were asked to provide assessments of a ten-minute release involving the largest container of a toxic chemical on their sites.[16] Some of these hypothetical cases included a train car filled with 180,000 pounds of chlorine, and sulfur-based compounds spreading from a manufacturing site to populated areas.[17] When it came time to fulfill this mandate in 1999, however, the US Environmental Protection Agency (EPA), the FBI, and the US Congress finally agreed not to publish this kind of information, at least not yet. The US government was concerned that terrorists could exploit "worst-case" release data on 66,000 US chemical facilities nationwide.

More recently, following the disclosure that a small but significant number of individuals were being granted so-called "Hazmat licenses" either illegally or with insufficient training, the federal government has ordered much closer scrutiny of how drivers are trained, licensed, and monitored. There are in the US approximately 80,000 companies that haul hazardous materials, and they make on average 1.2 million shipments every day. What they carry ranges from the relatively benign to the extremely dangerous, including medical waste, highly toxic chemicals, and extremely flammable liquids and gases. The transport and handling of such materials are are federally regulated, but the drivers are licensed by states. Typically, drivers in this industry must have both a commercial driver's license and what is called a hazardous materials "endorsement." Under new rules proposed by the Federal Motor Carrier Safety Administration, an agency of the US Department of Transportation

(DOT) and formerly part of the Federal Highway Administration, the simple four-hour course required to get the endorsement may be supplemented with additional training, and by an act of Congress DOT is being asked to implement background checks on any individual applying for or renewing a license to transport these materials.

Beyond active monitoring and surveillance—at weigh stations, and within and at the perimeters of chemical plants and refineries—vigilant security is still the most practical form of defense. Technologies to monitor the presence of CW agents, whiled further advanced than those under development to monitor biological agents, are still not practical for widespread installation. During the Gulf War, a number of CW sensors were deployed in order to alert troops to the start of a CW attack. They included fairly simple devices, such as chemically sensitive paper bracelets that would change color in the presence of certain agents. More complex technologies, from portable chemical kits to automatic electronic alarms and even mass spectrometer devices (which perform chemical analysis by ionizing gases) were also deployed, but with limited success. The challenge with these technologies was roughly the same (although not as limiting) as what we have discussed in terms of sensors for biological agents. First, they all have a problem with sorting out "background noise." That is, they have trouble discriminating between relatively harmless airborne substances (smoke, or petroleum fumes) and actual CW agents, and therefore produce false positives. And second, they all perform less than ideally in terms of speed, and in a chemical attack, speed of detection is obviously of critical importance. While advanced research projects on CW detection equipment continues in the US and other militaries, and includes even hand-held and pocket-sized devices, for the general public, human monitoring and surveillance are still the most effective responses to CW threats.

Bhopal

In considering CW attacks, it may be instructive to look in some detail at a particularly devastating instance of chemical poisoning even though it was not, strictly speaking, a terrorist attack, since the intention of the probable perpetrator was certainly not to cause death, but to commit an act of corporate sabotage. Nonetheless, deaths did occur in this incident at a chemical plant, and on a very large scale. The FBI has long warned that an attack on an industrial facility would most likely come from "a local thug who already knows the chemicals are there or from a disgruntled employee."[18] This was the situation in the December 1984 release of a toxic chemical in Bhopal, India. The largest industrial catastrophe to date—some details of which are still unclear to this day—began when a massive release of methyl isocyanate (MIC) killed at least 2500 people. Thousands more suffered permanent injuries because of damage to the respiratory system.[19]

The factory in question was owned by Union Carbide of India Limited, and it manufactured Sevin, a carbamate pesticide, in Bhopal, the capitol of Madhya Pradesh. At approximately 12:30 A.M. on December 3, 1984, thousands of pounds of MIC escaped into the air and the shantytowns surrounding the plant. An inversion layer had formed following a cool evening, and a slow but steady wind carried the heavier-than-air chemical toward populated areas several kilometers to the south, increasing the lethality of the event.[20] The civil case that followed on behalf of the victims claimed that unsafe engineering standards and the lack of attention by the American corporate parent Union Carbide were at fault. After years of legal and political wrangling on the part of the Indian Government and US trial lawyers, Union Carbide agreed to a $470 million settlement for the victims.

According to the Indian Government, the Bhopal disaster occurred after a Union Carbide worker, in the normal process of cleaning pipes, accidentally mixed water with MIC because a safety device was not in place. The mixture of water with the MIC then caused a violent reaction in one of three large-capacity tanks.[21] Interestingly, just hours after the accident, it was even suggested that the disaster was orchestrated by Sikh terrorists.[22] However, Union Carbide investigators and a study performed on its behalf by Arthur D. Little, Inc., suggest another explanation that points to deliberate malfeasance. A dozen years after having published his results, Ashkok Kalelkar, the author of the Little study, is "absolutely convinced that it was a wanton act of sabotage, and only sabotage, that caused the Bhopal disaster."[23] In this explanation of the incident, the trouble started when a worker who'd been demoted decided to get back at the company by ruining a batch of MIC. (He knew, from company manuals, that water and MIC should not be mixed.) Almost certainly not cognizant of the huge catastrophe that would result—after all, his own family lived near the plant—he connected a water hose directly to tank, having first unscrewed a meter cap on the tank, and let the water flow in. When other employees at the plant discovered something was wrong, they made a desperate attempt to solve the problem, but the contents of the tank were drawn out toward the processing station, which resulted in the sudden release of MIC.[24]

Certainly in the tens of thousands of chemical plants and storage facilities in the US, there are any number of employees who might have access to controls that when abused, inadvertently or deliberately, could cause tremendous chemical catastrophes. What we as a society have to rely on at a minimum is monitoring, surveillance, and workplace management and supervision, not to mention faith in our co-workers and fellow citizens.

Densely Populated Spaces

Large numbers of civilian casualties could result should terrorists employ chemical or biological agents at densely populated and especially enclosed locales, such as commuter trains, indoor arenas, theaters, malls, and even in tunnels. In 2000, tests were performed by various US government agencies to examine how aerosols or vapors from toxic compounds would behave in US commuter subways. Not surprisingly, the tests concluded that the effects of a CBW attack in an enclosed or semi-enclosed space are likely to be more serious than the effects in an outdoor attack. That's not to say, however, that an enclosed space is always a bad place to be.

Most buildings, in the event of a massive outdoor attack, would provide at least a modicum of protection from CW and BW agents. In such a situation, a large modern building's ventilation system could be set so that air pressure within the building would be slightly higher than the pressure outside, causing the flow of air to go from inside to outside. As long as the intake ports of the system were filtered, and not simply pulling in tainted outside air, such a "positive pressure" procedure could go some way toward protecting inhabitants. (In most modern office buildings, 20 to 25 percent of the air in any given ventilation cycle comes from the outdoor atmosphere; the rest is recirculated.)

Other HVAC (heating, ventilation, and air conditioning) modifications could be put in place in advance of a CBW incident. For one thing, buildings' HVAC systems could (and probably should) be designed with so-called "zone isolation" features, making it possible to segregate systems of ducts and blowers so that an interior attack with an agent could be prevented from spreading through the entire structure. Of course to make such a safeguard effective, building engineers would need to be immediately alerted to any CBW incident that took place within the structure.

More advanced "immune building" studies and prototypes are being run by DARPA on behalf of the US military, and include technological solutions such as high-grade filters, systems that employ high-powered ultraviolet rays to decontaminate airflows, and automated detection and response systems to isolate affected sectors of a structure or virtually eliminate the intake of outdoor air. But again, for most people, it is the basics of building safety that will make a difference during an attack—being familiar with emergency evacuation routes and regularly checking that they are clear and lighted and otherwise in working order; knowing which co-workers are charged with leading and directing personnel in an emergency, and which personnel may need special help; and conducting frequent and serious drills on how to respond in the event of an emergency. These are routine and common-sense measures, but in a crisis, they will pay off.

Water Supplies

Earlier in this chapter we quoted the terrorist Ramsi Yousef in a statement he made about using chemicals as weapons ". . . for use against vital institutions and residential populations and drinking water sources and others. . . ." A prime source of concern in any overview of CBW terrorism must be the US water supply, a complex system that feeds the nation's thirst for water. Thousands of companies and utilities bring water to homes, businesses, and industry. The New York City water system alone, for example, supplies 1.3 billion gallons per day to its customers. In California, where a complex man-made water system made possible the very existence of Los Angeles, the water supply now serves 3.8 million residents.

Water systems are just that—truly *systems*, complex organizations that rely on the close interworking of many components. The California State Water Program that serves 29 urban and agricultural water suppliers, for example, encompasses 32 storage facilities, reservoirs, and lakes, 17 pumping plants, 5 hydroelectric power plants, and 660 miles of open canals and pipelines. Wells also figure in the water supply. New Jersey alone, in addition to above-ground sources, uses public wells to serve about 2 million people (out of a population of some 8.5 million). In all, there are 168,000 public water systems in the United States. Of these, about 3000 metropolitan systems supply water to approximately 70 percent of the nation's population. Such a huge and sprawling structure is obviously vital to the country's well being, and because of its importance, the number of parts necessary for it to operate, and their far-flung nature, it seems a likely target for attack by terrorists.

In imagining a CBW attack it is probably easy to conjure visions of black-clad figures creeping through the woods surrounding secluded reservoirs. Could these shadowy figures then pour vials of biotoxins into the water and slink away, leaving the deadly materials to disperse and flow into city water systems, threatening all its users? Actually, the dangers posed by such a scenario are small. The FBI's Deputy Assistant Director in the Counter Terrorism Division said this in a report before Congress:

> . . . Affecting a city-sized population by a hazardous industrial chemical attack on a drinking water supply is not credible. A hazardous industrial chemical attack on a post-purification drinking water storage facility in a small municipality or a building-specific target is likely to be more credible but difficult to carry out without site-specific knowledge and access.[25]

Most professionals who have looked at a direct attack on drinking water dismiss its likelihood. "The water threat is mostly science fiction," said one microbiologist who formerly led the UN's biological weapons inspection teams in Iraq.[26] For example, suppose terrorists wanted to use botulinum toxin as their poison of choice. Despite its high toxicity, the toxin would be so diluted by millions of gallons of water that it would be ineffective. Just as importantly, many toxins, and botulinum is no exception, are quire fragile and easily rendered harmless in chlorinated water.

Poisons put directly into reservoirs are impractical. The amount of a chemical or biological toxin needed to harm people would be immense. "You're talking 18-wheelers full of things and trying to somehow dump this stuff in with nobody noticing them," said Jeff Danneels, at the Sandia National Laboratories in New Mexico. "It's an enormous task if you sit down and do the calculations."[27] Large amounts of harmful chemicals added directly to the water of reservoirs would have immediately noticeable effects. Some poisons would color the reservoir's water or leave a distinct smell. And fish killed by the poisons would attract attention long before tainted water made its way through pipes and filtration plants to the people who used it. (In some larger systems, it takes months if not more than a year for a given volume of water to run the course from reservoir to tap.)

In addition, already existing water-purification methods such as chlorination and ozonation would be effective against most toxins added directly to water. Cities and utilities also test water regularly for the presence of dozens of substances in their water supplies. "People have to remember that normal contaminant barriers that we use day to day, year to year, go a long way to protect the customer," said a spokesman for a New Jersey water company. "We have tremendous power in terms of cleaning and purifying water."[28]

If testing revealed a noxious substance, many cities could close off the tainted source and divert to other supplies. New York City, for example, has about 20 reservoirs to draw on. Cities and utilities have also increased testing efforts for their water because of perceived threats. Some companies and systems are additionally considering the use of robot sensors in the water pipes that would warn of the presence of deadly materials that have been placed in the water. Nevertheless, certain newer water purification methods may have added to risks of contamination by terrorists. Some water systems have shifted away from highly reactive and dangerous chlorine to milder detergents, but these in turn might not be as effective as chlorine in combating materials that terrorists would use.

Also, is worth pointing out that some compounds, such as sarin or the very toxic fluoroacetates, can survive for some lengths of time even in modern water systems. Still, terrorists would require large amounts of them and knowledge of where exactly to insert the toxins. Even deadly toxins such as botulinum would be diluted past effectiveness, as well as weakened by chemical treatments and environmental factors. It is likely that protein-based toxins would be rapidly denatured and lose their toxic effects. Other compounds, such as nerve agents, in addition to becoming less and less concentrated as they mix with water flows, slowly hydrolyze and break apart, effectively rendering them nontoxic.

Cautionary tales do exist, however, and public water supplies warrant vigilance even without a serious terrorist threat. The 1993 outbreak of *Cryptosporidium parvum*, a protozoan parasitic organism, affected some 400,000 people in Milwaukee, Wisconsin.[29] Although those who had immune deficiency syndromes from various causes were most seriously affected, most recovered from the gastrointestinal disease. We are only now beginning to better understand what exactly happened, but it was almost certainly a man-made disaster, caused by sewage that spilled out into a river and subsequently drawn into the water system. Terrorists could, conceivably, deliberately contaminate water sources in this way.

Although attempts to poison a water source are unlikely to succeed, state governments and utility companies have mounted campaigns to protect reservoirs. Many places have cut access to reservoirs. The Metropolitan District Commission in Massachusetts, for example, has canceled boating and fishing programs and denied the public access to lands and roads at the huge Quabbin Reservoir, a key part of the Boston area's water supply. New York City, too, has closed roads running around reservoirs and over dams in its upstate watershed, and Colorado has barred or severely restricted boating at a dozen government-run bodies of water. Other states have taken similar measures.

Threats to the water supply also exist nearer to the point of delivery. Cities have thousands of miles of pipe beneath their streets through which water flows. CBW terrorists could patch into lines to neighborhoods after water has been treated and checked for safety at a central location. They could then "back-siphon" chemicals or biotoxins into the water supply of a building or neighborhood. Such local water lines have not been a major security concern in the past, but the perception of the threat they pose has changed. "If someone is going to attack us, that's where they would do it," said Dr. Dennis D. Juranek, associate director of the Division of Parasitic Diseases at the Centers for Disease Control and Prevention. "We're highly vulnerable."[30] The FBI agrees, saying a "successful attack would likely involve either disruption of the water treatment process (for example, destruction of plumbing or release of disinfectants) or post-treatment contamination near the target."[31] If terrorists pulled off such an attack, authorities could react promptly to seal off affected neighborhoods and buildings. The resulting loss of life would be limited, and not on the scale that the terrorists might wish.

In response to this infrastructure threat, some cities are taking additional preventive measures. They are attempting to limit access to tunnels and water mains and adding new locks and barriers to these structures. Use of surveillance cameras at key locations is growing, as is the use of alarms to alert authorities to intrusions into the water system.

Government and water utility officials prefer not to be specific in their discussions of the new security measures, of course. "We don't need to advertise where the weakest links in the armor are," said Tom Curtis, Deputy Director of the American Water Works Association.[32] The officials do have confidence in their responses to meet any threats that may arise: "People have to remember that normal contaminant barriers that we use day-to-day, year-to-year, go a long way to protect the customer," said United Water spokesman Kevin Doell. "We have tremendous power in terms of cleaning and purifying water."[33]

The federal government, meanwhile, has stepped up its role in trying to guarantee water safety. Environmental Protection Agency administrator Christine Todd Whitman announced shortly after the World Trade Center collapse the establishment of a water protection task force that will aid federal, state, and local bodies in strengthening their ability to safeguard the nation's drinking water supply from terrorist attack. She reminded the nation that "the threat of public harm from an attack on our nation's water supply is small. Our goal here is to ensure that drinking water utilities in every community have access to the best scientific information and technical expertise they need and to know what immediate steps to take and to whom to turn to for help."[34]

Agriculture

Probably one of the more alarming scenarios put forth after the September 11 attacks was not against our agricultural system, but instead one that took advantage of it. When it was learned that the terrorists had looked into flying crop-duster aircraft, the Federal Aviation Administration (FAA) mandated a short moratorium on crop-duster flights. At first glance, it would seem that most crop sprayers would make excellent delivery devices for biological agents or very toxic compounds such as VX. But the relatively low volume carried by such craft and the typical design of their spray attachments make them not really suitable for such purposes.

Standard crop dusting involves the dispersal of pesticides (and insecticides, herbicides, fungicides, and so forth) in relatively large particle sizes. For example, spraying for gypsy moths from 50 feet usually requires particle diameters of about 150 microns in order for the agent to stay "on target." (Any smaller insecticide particles from that altitude would be likely to drift, and that would be a real economic and regulatory problem for any crop-dusting company.) Drops of this size are more than 10 times larger, for example, than the nominal particle sizes required for BW agent delivery. So could such equipment be used for biological warfare agents? In modern crop-dusting operations, sprayers that would suit standard BW agents—which generally have particle sizes between 1 and 10 microns—are not widely available for aircraft. To make use of a plane, then, the terrorist would most likely have to find BW agents in the form of a liquid suspension, and this would not be easy.

The case of CW agents, however, is probably a different story. VX, the deadly nerve agent, could in all likelihood be dispersed from a crop duster in a relatively normal configuration. Because VX is essentially a "contact" agent, doing its work by landing on the skin, and not needing to be aerosolized into a form that can be inhaled, larger droplets might work. It is possible that they might fall where the pilot wanted, even from significant altitudes.

But that is not to say that dissemination of VX from such a plane would be easy. The main problem would be in the actual loading of toxic compounds into the spray mechanism. Would the seals be good enough to prevent leakage? Could the pilot be adequately protected? The technique in such an operation would need to be perfect, since even minimal exposure to VX could kill the pilot even before the aircraft left the ground. Furthermore, to actually accomplish such a mission, the potential terrorists would have to hire or infiltrate not just a flying company, but a separate company to do the actual loading of the chemicals, since those two functions are always performed by separate specialized firms—those that fill the plane's tanks, and those that supply the pilots who do the flying and the discharge. So it is reasonable to say, especially given current circumstances, that such an attack is a highly unlikely event, even if the potential attackers could get past the surveillance procedures, background checks, and other security reviews any reputable company is certain to put in their way.

The deliberate use of biological agents against ongoing agricultural enterprises is another potential threat that has to be considered. While farming has not in the past been of great interest to terrorists, the consolidation and mechanization that's gone on in modern agribusiness could make food-growing operations more attractive targets. For one thing, large farms could present opportunities to set in motion outbreaks of disease among livestock. Foot-and-mouth disease, a viral infection that affects most domesticated animals except horses, will probably cost the United Kingdom an estimated $3.5 billion.[35] And we have already seen the ongoing problems to the economy and consumer confidence caused by Bovine Spongiform Encephalitis (BSE, or Mad Cow disease) in England and elsewhere.

Other food security problems could be presented by the willful adulteration of foodstuffs using chemical or biological agents. While we have no more than isolated cases of criminal activity in this regard (recall the Tylenol cyanide case in the 1980s), there are enough examples of accidental cases of food poisoning or food-borne infection to give us pause. In Japan, for example, nearly 9000 people were eventually affected by a single incident of pathogenic *E. coli* bacteria in 1996. Two years later, also in Japan, 10,000 people were sickened by powdered milk contaminated with staphylococcal enterotoxin. Could perpetrators cause similar outbreaks by contaminating food supplies? The probabilities are difficult to ascertain, but it may be wise to increase security through better awareness of how food is grown, processed, and delivered to consumers, not just in the United States but around the world.

LIVING WITH UNCERTAINTY

As we have said before, an important part of the fear that chemical and biological warfare inspires grows out of the fact that we know so little about it. To many people it seems to represent the sinister unknown. Almost no one, at least among regular citizens, civilian health professionals, and even national political leaders, seems to have had prior first-hand experience of it. And it is understandable that many of us feel beleaguered, unmoored, now that we are beginning to see it in our own midst. But we can at least take some of the uncertainty out of our lives. For one thing, we can learn more about these weapons. The truth is, CBW agents have been around for a long time, and there is a body of knowledge, and a community of scientists, most of them from the military, who have spent years working with, or just as often against, the development and deployment of these weapons.

And we can also, to reduce uncertainty even further, take steps to secure our own society and our personal lives. All of us would like, to be sure, a simple quick fix to this situation. That's understandable. But the process of making ourselves more secure will not be simple. To be effective, security requires human intelligence. It can't rely solely on even the most advanced mechanical or electronic or any other kind of device to detect threats and remove them from our lives. It can't rely on software that automatically traps and counters invasive attacks. There is no computer or video surveillance or chemical or biological monitoring device that can, in and of itself, be substituted for informed and intelligent human watchfulness, analysis, and judgment.

PART TWO

Chemical Agents

CHAPTER 4

Basic Concepts

All substances are poisons. There is none which
is not a poison. The right dose differentiates a
poison from a remedy.

—*Paracelsus 1492–1541*[1]

Chemical weapons utilize the inherent toxic properties of a substance to poison—by inhalation, ingestion, contact with the skin, or a combination of all three. Poisoning as a method of assassination is no doubt older than recorded history, and history is replete with examples of noxious smokes and gases being used in battle. And certainly there were instances in the distant past when poisons were used against large populations. But it was not until the twentieth century that armaments makers perfected the means to manufacture large quantities of chemicals and systems capable of delivering them over great distances and on a massive scale. Modern artillery, including cannon, rockets, long-range ballistic missiles, and even unmanned aerial vehicles can now be armed with highly toxic chemicals. And when chemical agents are combined with such sophisticated delivery systems, they surely qualify as and are intended as weapons of mass destruction.

As recent experience has shown, however, it is dangerous to underestimate the resourcefulness of a determined adversary. While the wide and effective dispersal of CW agents generally requires some sort of sophisticated delivery system, and while some of the public's understandable alarm is exaggerated, a crude apparatus, even a crop-dusting plane or simple canister in the hands of a fanatical subway rider, could cause a devastating amount of injury.

WHAT CHEMICAL WEAPONS ARE NOT

Maybe the easiest way to begin thinking about chemical weapons is to compare them to other types of armaments—to distinguish them from what they are *not*. Chemical weapons are in the first place different from *kinetic weapons*, such as bullets, other projectiles, and shrapnel, which create casualties using force of impact. The lethality of a kinetic weapon depends on its size and its force at impact, so dense materials like steel and lead (and, more recently, the even denser metal uranium, in depleted form) are the chosen materials.

Incendiaries are armaments that act primarily by burning and creating large amounts of concentrated heat, and include thickened gasoline preparations such as napalm (gelatinized fuel, ignited by white phosphorus) and fuel-air munitions using ethylene oxide, which create large airborne "firestorms." Although some incendiary devices produce shards of white phosphorus that can act as a poison when fragmented and lodged in tissue, they are not generally considered CW agents.[2]

Radiating (nuclear) weapons produce energy in the form of an explosive blast, in addition to gamma rays and neutrons that destroy unprotected tissue, particularly DNA. (Thus, mustard agents and T2 mycotoxin, because of their similar effects, are sometimes referred to as "radiomimetic.") Enhanced radiation warheads, or "neutron bombs," minimize the destruction of materials while maximizing lethalities among enemy personnel.

Last are *biological weapons*, which utilize living organisms, or toxins generated by living organisms, to cause death or incapacitation. Sometimes the lines between CW and BW can be blurred. For example, the Chemical Weapons Convention (CWC), which was adopted by most nations in 1993 and prohibits the development, use, or possession of any chemical to be used as a weapon, lists ricin (a toxic substance that is extracted from the castor bean plant) as a chemical weapon, and in a broad sense any toxin from an organism is still essentially a chemical. Still, for the purposes of this book, we will use the BW label to refer to agents that use living organisms and their toxins in order to inflict injury and death.

WHAT CHEMICAL WEAPONS ARE

Chemical weapons, then, are those that deliver poisonous substances into a target population, with the purpose of causing injury, incapacity, or death. Not surprisingly, historians, ethicists, and students of military science do not all agree on a single strict definition. The rather extensive literature on CW agents includes not only the usual suspects (blister, nerve, and choking agents), but napalm, tear gas, smokes, and herbicides/defoliants. In this book, we tend toward the "usual suspects" definition.

Properties

CW agents can be categorized based on their physical and chemical properties, as well as their toxicity and mechanisms of injury. First and foremost, a chemical can be defined in terms of its *lethality*—how likely it is to result in the target's death by poisoning. There is not always, as it turns out, a clear distinction between lethal and nonlethal agents. Lethal CW agents are designed of course to cause fatalities under battlefield conditions, but sublethal doses can be employed to injure or incapacitate rather than kill. For example, at one time the United States military employed a chemical doctrine that called for shelling a given target in such a way as to generate half the lethal concentrations of sarin necessary to cause death, thereby mostly incapacitating the enemy rather than killing them outright. This was done not out of humane impulses but as a way of conserving ammunition. Nonlethal riot-control agents, on the other hand, while designed primarily to incapacitate, can also cause death. The lethal dose of the common tear gas CS (ortho-chlorobenzylidene malononitrile), for example, is approximately two *ounces* if swallowed, an amount that as a practical matter would be difficult to ingest all at once. Death or permanent injury from CS tear gas, however, could also occur if this agent is used in enclosed spaces such as tunnels.

It is also important to consider a CW agent's *mode of action*, which indicates the route the toxic chemical takes into the target's body. The most common routes of exposure are inhalational (via the respiratory tract) and percutaneous (through the skin). An agent that acts via inhalation damages the lungs, or at the least passes rapidly into the bloodstream, where it does its work. An agent that acts percutaneously enters the body through the skin, eyes, or mucous membranes and can also reach the bloodstream. Less useful on the battlefield but still valid for terrorist purposes are poisons that can be administered orally (for example, by contaminating food, tampering with over-the-counter medications, or poisoning drinking water).

Another metric used to classify chemical agents is *speed of action*, a measure of the delay between exposure and effect. Fast-acting poisons, such as nerve agents and cyanide, can cause symptoms to appear almost instantaneously and might cause fatalities in as little as a few minutes. Slower-acting agents like mustard can, depending on the amount of exposure, take hours to effect serious injury.

Related to lethality is the CW property called *toxicity*, which is a measure of the quantity of a substance required to achieve a specific, deleterious effect. CW agents are essentially toxic compounds that can poison via inhalation or skin contact (or both). For example, 3200 milligrams (or 3.2 grams) per cubic meter of air of the World War I choking agent phosgene will kill at least half of the personnel breathing the gas. Comparatively, only 70 milligrams of the nerve agent sarin is required to achieve the same effect. In other words, sarin is almost 50 times more toxic than phosgene. All things being equal, for a substance to make an effective CW agent its toxicity must be extremely high, and its effects more or less immediate (within minutes to hours).

CW agents are also measured along a continuum of *persistency*, which indicates the length of time a CW agent remains a hazard after its release. Nonpersistent agents tend to be rather volatile and to evaporate quickly, dissipating within a few minutes to about an hour. German military tacticians in World War I, for example, estimated that 12,000 kg of non-persistent agent munitions were required to cover one square kilometer on the battlefield in order to be effective. Semi-persistent agents generally linger for several hours to one day. Persistent agents, which tend to be rather thick and oily, can last for several days to a few weeks, depending on ambient temperature and other conditions.

CW agents are also categorized by their *state*—their physical form may be solid, liquid, or gaseous. Most are stored and delivered in liquid form, including "nerve gas," "mustard gas," and "poison gas"—all of which are technically aerosolized liquids. (This misnomer has its roots in the first use of chemicals, some of which were in true gaseous form, in World War I.) There is no reason why a CW agent must take on any particular state to be effective, but engineering the agent into a practical weapon—that is, *weaponizing* the agent—may be more or less of a challenge depending on whether the source chemical is a solid, liquid, or gas at room temperature.

Delivery Systems

Any discussion of chemical or biological weaponry must address the importance of aerosols for the delivery of toxic agents. (The very term "aerosol" was coined during World War I by F. G. Donnan, who was attempting to characterize the behavior of toxic smokes.) In terms of military efficiency, CW systems require substantial generation of aerosols—tiny liquid or solid particles suspended in air—to create the concentrations necessary to generate sufficient casualties. (We will have much more to say about aerosols and their behaviors in Part III, on biological weapons. For now, it is only necessary to treat this subject in brief.)

As mentioned earlier, inhaling CW agents is one of the most efficient means of creating casualties on the battlefield or in a terrorist attack. In order to maintain high concentrations of a given CW agent in a targeted area, the particles generated have to be small enough to deposit in the lungs (especially down to the alveoli of the lung, where gas exchange takes place) or in the upper respiratory system. However, they should not be so small that they simply are inhaled and expired all in the same breath cycle. On the other hand, should the particles be too large, they rapidly fall out of the air and land on the ground, with fewer chances of injuring personnel (again, depending upon the type of CW agent used).

Particles with diameters in the 1- to 5-micron range are optimum for lung deposition.[3] Especially in the case of CW agents, particles of 10 microns or larger may not reach the alveoli, but making contact with the respiratory mucosal linings will still bring about intoxication. Much larger particles (say, of 70 microns) fall out of air suspension quite quickly, and are much less likely to be inhaled (especially if some kind of mask is being used). Even so, such large particles may be an effective form of delivery if contact on skin or equipment is the desired goal (VX is a good example).

The first CW agent delivery system was fairly simple. In World War I, cylinders of liquefied chlorine were brought to the front, and the agent was released in the form of a gas. The prevailing wind did the rest of the work, bringing the chlorine to the enemy trenches. We are not likely to see this type of delivery system in modern warfare, although terrorists could improvise something similar on smaller scales.

The most efficient form of delivery developed in World War I was the Livens projector, which was basically a large mortar round (artillery shell) containing a CW agent or its components. The chemicals were stored in a large cylinder with an explosive charge running through the center. When the cylinder was lobbed toward the enemy it exploded, and a mixture of aerosol, larger airborne droplets, and toxic vapors were thus created. Depending on the circumstances, the more efficient shells are those that explode at a given calculated height. Detonating the shell as it neared its target produces a fair amount of aerosol, and a range of droplets from small to large depending on the engineering of the munition.

Various militaries and CW experts may differ as to how much CW agent survives the thermal destructive effects of the explosive used in delivery shells, but generally speaking loss of CW agent from detonation is not very significant. In the case of bromobenzyl cyanide (CA), for example, World War I munitions lost up to 25 percent or more, but later improvements have reduced CW agent loss considerably. An East German source on CW reports that chemical agent efficiency has since increased to 95 or 99 percent,[4] and not the much lower figures sometimes seen in other references (such as a Chinese book on CW that claims US munitions were only 75 percent efficient in the case of VX agent in a 155-mm shell[5]).

Weaponization of CW agents usually involves addition of other compounds (sometimes other CW agents) in order to stabilize the contents, or to increase the range of temperatures in which the chemical agents would be effective. Thus, rather mundane issues of storage and handling of chemical weapons are actually very important problems to consider.

BASIC CLASSES OF CW AGENTS

The following are chemical weapons that have been used or stockpiled for use in war. This is not an exhaustive list, but represents the basic classes of chemical armaments that have proliferated since World War I.

- Choking gases (lung irritants)
- Blister agents (vesicants)
- Blood agents
- Nerve agents (toxic organophosphates)
- Incapacitants (psychoactive chemicals)
- Harassing or riot-control agents (RCAs)
- Vomiting agents

Most modern military and terrorist experts believe that the most important CW agents are in the the nerve and blister classes. Both have proven themselves to be effective, are liquid at a wide range of temperatures, and have a long record of research and development.

If a modern military is looking for a low-cost, effective chemical weapon that can be mass-produced with relative ease, then mustard and Lewisite are the logical choices. These agents are capable of causing widespread injury to the enemy. While few of those affected would actually die from a mustard or Lewisite attack, the injuries suffered would cause the enemy great distress and put enormous demands on its medical and logistical system.

Organophosphates (nerve agents) require more sophistication but they too are not outside the reach of most countries, as Iraq demonstrated in the 1980s and 1990s. Terrorists probably would try to acquire the highly toxic nerve agents, which have the potential to create many casualties with great lethality and require much smaller quantities than mustard or Lewisite.

In any event, if you remember two representatives from each of these classes—the blister agents mustard and Lewisite, and the nerve agents sarin and VX—you will have in memory a reasonably clear list of the most likely chemical threats, both military and from possible use in terrorism. Other CW agents—an extensive list of them will follow shortly—are of course potential threats, as described below, but mustard, Lewisite, sarin, and VX—as well as we can tell right now—are the modern poisons of choice.

Choking Gases (Lung Irritants)

A brief and somewhat grim quotation does enough to summarize the effect of these CW agents.

Choking agents cause injury chiefly in the respiratory tract—that is, in the nose, throat, and, particularly, the lungs. In extreme cases membranes swell, the lungs become filled with liquid, and death results from lack of oxygen; thus, these agents "choke" an unprotected man. Fatalities of this type are referred to as "dry-land drownings."

—*US Army Field Manual, FM 3-9*[6]

Choking agents were among the first CW armaments produced in large quantities, and were used extensively during World War I. Chlorine was used in the first major attack, its attraction being that even as a gas it is heavier than air, and thus particularly well suited to trench warfare.

Of the choking agents, chlorine and phosgene are the best known. Inhaled in sufficient quantities, they induce pulmonary edema, ultimately suffocating the victim in his own fluids. Both chlorine and phosgene are used in many chemical industrial processes on the order of millions of tons annually,[7] making the control of these compounds problematic.[8] Chloropicrin (or chlorpicrin) is also extremely irritating to the eyes, nose, and throat, as well as having lung-irritating effects.

The technology and knowledge used to make these first-generation agents (or asphyxiants) are widely available, but for several reasons they are now less attractive for military operations. First, many have a strong odor (less so in the case of phosgene) and create irritation that alerts the victim to their presence, allowing time to don protective gear or vacate the affected area. Second, compounds such as chlorine and phosgene require relatively high concentrations to reach lethal levels. Finally, and related to the last, the utility of choking agents in modern warfare is reduced by their tendency to dissipate very quickly. (This is unlike the situation in World War I, when volatility—and the tendency to dissipate quickly—was a sought-after quality in CW agents used against the enemy.)

The Bhopal industrial disaster of 1984 is an object lesson in the hazardous nature of pulmonary irritants. In this case, an accidental release of methyl isocyanate—a compound used to manufacture polyurethane and pesticides—killed between 2500 and 5000 people, and injured tens of thousands more.[9] As mentioned earlier, much of the available evidence indicates that the incident was due to sabotage.[10]

Chlorine

At room temperature, chlorine (Cl_2) is a pungent, green-yellow gas; it can be liquefied under moderate pressure. In World War I, chlorine was brought to the front in the form of a liquid, one quart yielding over 100 gallons of chlorine gas upon release. For the conditions prevalent in World War I, chlorine satisfied some essential requirements for a chemical weapon, being sufficiently toxic (although not as toxic as many other CW agents) and, perhaps most importantly, abundant and relatively cheap. Even before the mass production of chlorine got underway, it sold for about a nickel a pound.[11]

As a poisonous gas, chlorine irritates the nasal passages, constricts the chest, and in larger amounts (approximately 2.5 milligrams per quart of air) causes death by asphyxiation. Although it was initially effective in World War I, chlorine's distinctive bleach smell—and eerie, greenish color—easily made its presence known, and the advance warning allowed for defensive preparations and tactical retreat.

Chlorine's high reactivity, however, also meant that soldiers quickly found that makeshift masks soaked in chemicals (sodium thiosulfate, glycerine, and even urine) all offered good protection. Six months after first being used in World War I, chlorine by itself no longer made a significant impact on the battlefield, although it remained an essential part of phosgene mixtures later on and was a critical component in the production process of highly lethal compounds that were to follow.

The use of chlorine as a weapon is now considered obsolete, but occasionally it does reappear in modern conflicts. According to an unconfirmed report, the Liberation Tigers of Tamil Eelam (LTTE) used chlorine in a 1990 attack against Sri Lankan forces, but did not cause serious casualties. A report in 1997 claimed that Muslims in the Bosnian city of Tuzla, the site of a significant industrial chemical facility, produced 120-mm chlorine-filled mortar rounds in anticipation of conflict with Serbian-led forces. It is not publicly known if any of these shells were used.

When chlorine gas encounters water, such as the moisture found in air or in the tissues of the throat, larynx, and lungs, it reacts to form hydrochloric acid and hypochlorous acid. (The latter accounts for the strong odor of bleach.) The irritation upon lung tissues caused by these compounds is what leads to injury or death. Depending upon the amount of the agent present and the length of the period of exposure, chlorine gas first causes spasms within the respiratory system, specifically at the bronchi, and immediately begins to choke the victim. Irritation of the mucosal linings within the respiratory tree can also be severe enough to slough off tissue, and these fragments can obstruct lower bronchial passages. Eventually, the gas continues its way down to the tiny air sacks (alveoli) in the lungs that are responsible for blood-gas exchange. Damage caused to these tissues allows fluid to accumulate in the lungs, and results in death by pulmonary edema. Signs of the latter condition can occur two to four hours following moderate exposure to chlorine gas, while death usually occurs within 24 hours after acute exposure. Survivors generally do not suffer long-term effects from a single, toxic exposure to chlorine gas. According to some historians of the World War I, veterans claiming long-term disability from gas inhalation during the war more often than not suffered from other factors, such as chronic tuberculosis or smoking cigarettes.[12]

In the modern era, few analysts would regard chlorine as a credible military threat. However, considering the large amounts of chlorine that are used and shipped for commercial industry, this chemical could be used as a devastating low-tech weapon in the hands of terrorists.

Phosgene

Like chlorine, phosgene is delivered as a gas. During World War I it was primarily filled in artillery shells (Livens projectors) rather than fixed cylinders, as was the case with chlorine. Even at toxic levels, phosgene gas (carbonyl dichloride) has little distinguishing odor, and usually kills its victims only after a considerable delay (up to 24 hours). In one instance during World War I, a soldier was given the responsibility of checking phosgene canisters, all the while unaware that one of these canisters had formed a small leak. This soldier died from phosgene exposure on the following day. In another example, a German prisoner of war boasted to his British captors that Allied gas was ineffective, and though he was himself gassed he expressed confidence that he was well onto his recovery. Nonetheless, 24 hours later the German soldier also died from phosgene inhalation. Phosgene was alone responsible for some 80 percent of those killed by chemicals in World War I,[13] and from 1915 to 1918 some 150,000 tons of phosgene were produced.[14]

It was once commonly believed that phosgene exposure led to the formation of acid in the lungs, which then destroyed tissue. However, this explanation is not adequate to explain how phosgene can do so much irreparable damage to the lungs in very small concentrations.[15] (Experiments with dogs, for example, showed that the toxicity of phosgene is 800 times that of inhaled hydrochloric acid.[16]) While a certain amount of acid is generated, phosgene destroys lung tissues through a much more complicated sequence of events. Because it reacts with vulnerable molecular groups within amino acids that make up enzymes,[17] one of the major consequences of exposure is the disruption of the delicate surface tension maintained among the alveoli. When the tension is broken, a breach opens, resulting in pulmonary edema. That the process involves a great many steps—including the immobilization of enzyme, gradual decreases in the surfactant that maintains surface tension, and the buildup of fluids in the lung—may explain why it takes so long for dramatic symptoms to appear. Finally, acidosis within the lungs does aggravate the situation, further irritating delicate tissue and causing the victim to breathe faster and inhale more of the toxic gas. Phosgene not only destroys lung tissue after several hours following exposure, but reported examination of World War I gas warfare victims found that phosgene also causes specific injury to the central nervous system.

While today's military analysts differ on an assessment of the phosgene threat in modern warfare, in the late 1990s US armed forces fielded new chemical agent detectors that are sensitive to the presence of phosgene. At least one CW expert has also raised the issue of phosgene being released from the burning of plastic polymers, and that these present an increased danger of smoke inhalation from fires in modern structures. While the production of hydrochloric acid from burning plastics is of concern, it appears that phosgene is probably not a very significant by-product of the combustion of PVC (polyvinyl chloride) or other chlorinated plastics (but see the cyanide compounds, under "Blood Agents," below).

Diphosgene

Being liquid at room temperature, diphosgene, another World War I agent, is generally easier to handle than phosgene, and it is more persistent than either chlorine or phosgene. The mechanism of injury or death from diphosgene closely mirrors that of phosgene. It was first used by the *Wehrmacht* in May 1916, probably in retribution for the French military's use of phosgene (*Surpalite*) a few months earlier. In one particular World War I attack, 100,000 diphosgene shells were fired in a single engagement near Verdun.

Like phosgene, total numbers of casualties from diphosgene in World War I were relatively few compared to mustard, but it was one of the most deadly CW agents used in shells during the war. There is also another analogue, triphosgene, that possesses the toxic and physical properties of diphosgene, but information on its actual use is scarce.

Chloropicrin (or Chlorpicrin)

Discovered in 1848, chloropicrin (trichloronitromethane) is a very useful commercial chemical, often employed in pesticide controls of various types. As a CW agent, chloropicrin was first used by Russia in World War I, and was eventually delivered in artillery shells and cylinders by all sides. (It was given the name *vomiting gas* by the British, *Aquinite* by the French, and *Klop* by the Germans.)

Chloropicrin is an oily liquid, its intoxication leading to pulmonary edema, and its relative stability in water allows it to affect other organs of the body as well. While chloropicrin was not as deadly as chlorine or phosgene, it once served as a rather vicious entrée for other chemicals when used simultaneously. During World War I, chloropicrin broke through the filters in gas masks of that era, forcing soldiers to take their masks off and leaving them vulnerable to simultaneous attack by other lethal compounds. The processes involved in the production of chloropicrin are relatively simple and inexpensive. Because of its strong odor, and being one-fourth less toxic than phosgene, however, chloropicrin has not received the kind of attention reserved for more potent CW agents.

Chloropicrin is an example of a CW agent with commercial (dual-use) applications. Because of its distinctive odor, it is currently used as an odor adjunct for certain pesticides: in order to ward off public entry until treatment is complete, pest control companies use it in conjunction with pesticides. In fact, pest-control services are required by law in many US states to use incorporate chloropicrin as a warning agent. It is also an effective soil fumigant for use in agriculture and may replace methyl bromide for this purpose.

Ethyldichlorarsine

During World War I, part of the challenge in finding an effective CW agent for offensive purposes was to create large enough concentrations of CW agent for effective battlefield use. Therefore, volatile compounds were favored in many instances, leading to the development of ethyldichlorarsine. It is an arsenical compound first prepared in the nineteenth century and later developed by World War I German military chemists.

Categorizing ethyldichlorarsine can be somewhat confusing, for it has multiple effects on the human system. Fast acting (compared to mustard or phosgene), ethyldichlorarsine (also called *Dick* by the Germans) was produced in anticipation of a military offensive planned for spring 1918, and was intended to support German infantry operations. Despite the serious efforts made to weaponize this compound, little data exist as to the number of casualties directly caused by it.

Perfluoroisobutylene (PFIB)

PFIB is an industrial gas, most often encountered today as a byproduct from overheating and during the production of Teflon® (polytetrafluoroethylene). PFIB has also found uses as an intermediate in some industrial processes, including etching for semiconductor fabrication. The cause of "polymer fume fever," PFIB has the potential to be an asphyxiating weapon, causing pulmonary edema even at very low concentrations.

Like phosgene, PFIB has a latency period between exposure and symptoms. While toxicity data are sparse, in humans this latency period is estimated between one and four hours before signs of pulmonary edema manifest themselves. Because of its high toxicity, and the fact that it could "break" protective filters used by military forces, some have speculated that the Soviet Union (and perhaps other countries) had in fact weaponized PFIB. It was therefore brought to the attention of the Geneva-based Conference of Disarmament in 1989 by the United Kingdom, and PFIB was subsequently entered as a restricted chemical in the CWC. PFIB, when produced as an off-gas from incendiary fires or other heat sources, is also a potential hazard, as militaries around the world often employ Teflon in fibers, tarpaulins, and other matériel. Although open sources are scanty in terms of its use as a weapon, the high toxicity profile of PFIB (about ten times as toxic as phosgene) and its wide availability put this compound as a Schedule 2 toxic substance in the CWC.

Blister Agents (Vesicants)

Among the most widely used and stockpiled of the CW agents are blister agents (or vesicants). Mustard agent and Lewisite are the best known. Other vesicants that have been developed for warfare include arsenical compounds such as the methyl-, ethyl-, and phenyl-dichlorarsines, and even more obscure agents such as phenyldibromarsine and dibromethyl sulfide. Ethyldichlorarsine is best categorized as a lung irritant, and little evidence exists that methyl-dichlorarsine was ever used in World War I. (For these reasons and others, we will address only one of these later, namely phenyldichlorarsine.)

Mustard (sulfur) was first tested in combat in 1917 by the Germans and has been used since in several conflicts, including widespread employment during the Iran-Iraq War (1980–1988). As with phosgene, the action of mustard is delayed, but mustard typically kills only a relatively small percentage of all casualties. Exposure to the agent in liquid or vapor form results in serious skin irritation, temporary and sometimes permanent blindness, as well as life-threatening damage to the (primarily upper) respiratory system. Lewisite never found use in World War I, and phenyldichlorarsine seemed to have been employed more as an afterthought, primarily for its solvent properties in diphenylchlorarsine mixtures (see "Vomiting Agents," below). Lewisite, however, was used by the Japanese military during its invasion of China (1937–1945).

Mustard (Sulfur)

Also called *Lost* (derived from the names of researchers *L*ommel and *Steink*opf) and *Yperite* (its French and Russian nomenclature), mustard had been synthesized for more than 90 years before the start of World War I by a number of chemists, including Despretz in 1822, although he didn't identify his discovery. Riche, then later Guthrie in England, repeated the work of Despretz in 1860. Guthrie described his newly found substance as "smelling like mustard, tasting like garlic, and causing blisters after contact with the skin."[18] Mustard's appearance on the battlefield in 1917 may have also inde-

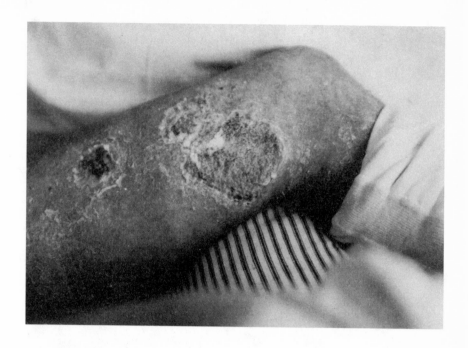

Injury from mustard, World War I vintage. Mustard injuries require a long time to fully heal. (Courtesy of Soldier Biological and Chemical Command, Historical Research and Response Team, Aberdeen Proving Ground, MD.)

A Japanese vesicant 150-mm mortar projectile, filled with vesicant (mustard or Lewisite), World War II. Note the identifying bands. (Courtesy of Soldier Biological and Chemical Command, Historical Research and Response Team, Aberdeen Proving Ground, MD.)

pendently given rise to being called mustard for the same reason. An oily liq-uid that is yellow-brown in its cruder preparations, sulfur mustard in its purer state is colorless and odorless. The first major attack using mustard came on July 12, 1917, in Flanders (near Ypres, Belgium, the site of the very first chem-ical attack in 1915). Germany fired 77-mm and 105-mm artillery shells filled with mustard against Allied troops. Aside from some sneezing among the exposed troops, no untoward reactions were noted. However, as the mustard vapors rose from their detonated shells and impact craters, in two hours or so, Allied soldiers complained of severe eye irritation, vomiting, and with signa-ture redness of skin and formation of blisters.

During World War I, mustard found its way through clothing, rubber, leather, and other protective garments in use at the time. Mustard's persistency and multi-route exposure hazard requires the defender to use full-body and cumbersome protective clothing, as well as time-consuming decontamination measures. As far as the German army was concerned, mustard was as much a defensive weapon, for it allowed the *Wehrmacht* to withstand repeated Allied offensives during World War I.

Despite the fact that it was widely described as a "gas," mustard is an oily liquid and is extremely toxic to unprotected skin, eyes, and the respiratory sys-tem. Because of its physical characteristics, it offered high persistency, creating extended periods during which targets suffered many injuries by secondary contamination. During the course of its action upon skin and other tissues, vesicles are formed after considerable delay (up to 24 hours in some cases), and these can turn into large and painful blisters. (This is why the term "vesicant" is used for CW agents of this type.) Blisters are formed by the agent's direct attack on cells. Mustard readily penetrates the skin and, being fat-soluble, attacks other organs of the body. Mustard literally destroys tissue from the inside. Initial exposure results in severe redness. When it was first used against the British, medics first thought they were dealing with an outbreak of scarlet fever.

Mustard has traditionally been described as having a "garlicky" odor,[19] but this is most likely due to either impurities or side-reactant products following detonation from explosive shells. Better chemical processes in the modern era would produce mustard with little or no odor.

Mustard is extremely poisonous, but not as lethal as many other CW agents. However, exposure to less than 1 gram of the agent in vapor form for 30 minutes will likely lead to death for an adult male. Most injuries (about 80 percent) from mustard during World War I were caused by having come into contact with mustard vapors.[20]

If a victim survives the initial encounter, mustard continues to destroy the body's own immune defenses and can complicate treatment of acquired infections. With large amounts of exposure, death from toxic shock can occur in about 48 hours. If mustard makes contact with the respiratory system, injury to tissue in the upper airways can cause necrotic tissue and pseudomembrane formations, and such casualties usually have poor outcomes.

Nitrogen Mustard

Nitrogen mustards are generally more toxic than the sulfur variety, and are also easily manufactured. Produced as a weapon in the 1920s and 1930s, nitrogen mustard was a by-product of Czech and German pharmaceutical work in quaternary ammonium compounds (or "quats"). Of the more common types of compounds that fall into this category is the quaternary ammonium compound used in household cleaners. Nitrogen mustards have also had medical applications, first for wart removal, and more recently in battling cancers.

Aside from being slightly less fat soluble, nitrogen mustards behave in much the same way as older sulfur mustards in terms of their toxicity, as well as possessing similar physical-chemical properties (oily liquids, yellow-brown color in cruder form). Another form of nitrogen mustard, HN-2, was probably developed as a spin-off from chemical-weapons work, but has found more peaceful applications as a chemotherapeutic agent. HN-2, known as mustine (Mustargen®)[21] or mechloroethamine, was once a widely used agent of its kind for chemotherapy. HN-3, on the other hand, has remained foremost as a CW agent, but is not easily stored for long periods of time. Being much more persistent than sulfur mustard (HD), HN-3 is considered militarily useful for contaminating enemy logistical supports, air fields, and terrain.

Lewisite

Because of its potential for proliferation, Lewisite is arguably one of the most important CW agents. First prepared in 1904, the "Dew of Death" was rediscovered in 1918 by the researcher from whom it got its name, W. Lee Lewis (1879–1943), of the Catholic University, Washington, DC. (The "Dew of Death" sobriquet was earned at least in part from the intended use of Lewisite by spraying it from aircraft.) In the later stages of World War I, a great effort was made to speed up production at a factory near Cleveland, Ohio, but the plant was completed only shortly before war's end. Approximately 150 tons of Lewisite were shipped to Europe by the time of the Armistice, but the agent was never used in battle, and the supplies were unceremoniously dumped into the ocean. The United States soon after closed down the Lewisite manufacturing facility, but France, Russia, Great Britain, and Japan stepped up production of the agent through the 1920s and 1930s.

The fact that Lewisite was non-flammable increased its favor by the US military later on. By the end of World War II, the United States had resumed

Lewisite production, but the vast majority of its output was also never deployed. After being neutralized with bleach, this batch of Lewisite was dumped in the Gulf of Mexico in 1946.

Lewisite in its purer form is an oily, colorless liquid, while impure Lewisite ranges from a light brown color to black. Its odor is often described as being reminiscent of geraniums, although purer distillates have very little smell. A less persistent liquid CW agent than mustard, Lewisite contains arsenic and is among several types of the so-called arsenicals that have been developed for warfare. Even in this group of compounds, Lewisite is notably toxic especially upon contact with the skin. Small amounts in liquid form on the skin causes pain within 12 seconds, and acid-burn like trauma begins between 5 and 15 minutes later, first forming smaller vesicles, then hours later blisters that resemble those that caused by sulfur mustard.

Although Lewisite has been sporadically used in the past, very little data from human exposure exist at this time. Consequently, much of what we know in the West about Lewisite's effects mostly comes from animal experimentation. Older data from the 1930s includes observing the effects of Lewisite on a human volunteer: In this case, Lewisite was completely absorbed in 5 minutes with a slight burning sensation, while mustard required from 20 to 30 minutes for absorption and produced no noticeable sensation. With Lewisite, the skin commences to redden at the end of 30 minutes; then the erythema increases and spreads rapidly. It occupies a surface of 12 by 15 centimeters toward the end of the third hour.[22]

While the toxicity of Lewisite is roughly the same as mustard, the action of Lewisite on the skin causes an immediate burning sensation, and its odor is readily apparent. Severe damage to the eyes occurs almost immediately after exposure, while Lewisite vapors irritate the mucosa of the nasal and upper respiratory system. Lewisite is subsequently absorbed into the body, and distributed as a systemic poison to various organs.

Historical evidence has pointed to use of Lewisite by Japanese Imperial troops against China during World War II. Ongoing assessments are being made of the extent of chemical weapons left behind on Chinese soil. An investigation led by a Japanese team in 1995 found traces of Lewisite at one site, as well as mustard and another arsenical agent, diphenylchloroarsine (see below, under "Phenyldichlorarsine").

Phosgene Oxime ("Nettle Gas")

In 1934, the chemical researcher J. Hackmann observed the effects of phosgene oxime, noting that "there are few substances in organic chemistry that exert such a violent effect on the human organism as this compound."[23] James Compton, an expert on CW agents and their effects, reports that phosgene oxime resulted from early Russian work in developing pesticides for cock-roaches.[24] During World War II, Hackmann and others worked with phosgene oxime ("Red Cross"), and probably other countries continued work in the nettle gases or urticants (itching agents) in their respective CW programs. Phosgene oximes and similar compounds are both damaging to the lungs while attacking the skin, and are sometimes referred to as *nettle gas* because its effects resemble those from the stinging nettle plant. There are actually at least nine analogues that make up the nettle gases, but phosgene oxime (dichloro-form oxime) is the most commonly cited version.

Not to be confused with the asphyxiant phosgene gas (see above, under "Choking Agents"), phosgene oxime is classified as a blister agent but has much faster action on the skin, causing painful sores that harden like bee stings. Irritation from phosgene oxime also extends to the eyes, and in higher concentrations phosgene oxime may cause pulmonary edema. It can be deliv-ered in powdered or aerosolized form.

Phosgene oxime probably attacks tissues by means of its chlorinating action with certain amino acid groups in proteins. While its effects are immediate, phosgene oxime also has been described as a long-lasting systemic poison with symptoms sometimes lasting up to a year. Some commentators have also compared the toxic nature of phosgene oxime to poisons found in jellyfish.

Because simple barriers can be effective in preventing skin exposure, the very reactive nature of phosgene oxime makes it less useful by itself as a chem-ical weapon, although it could be very effective when used against troops that have no CW agent defense equipment or training. Some CW preparations have employed a synergistic mixture of phosgene oxime with mustard, the idea being to facilitate openings for more efficient entry of mustard agent.

Phenyldichlorarsine (PD)

In the mid-1800s, Bayer, Dehn, and others synthesized compounds such as methyldichlorarsine, phenyldichlorarsine, and other related compounds, not-ing their potent irritating and toxic properties. Later during World War I, Fritz Haber directed the development of these as CW agents, including phenyl-dichlorarsine, at the German Kaiser Wilhelm Institute.

Phenyldichlorarsine proved to be a persistent compound. Although origi-nally introduced as a vesicant in World War I, phenyldichlorarsine also has very powerful irritating effects on the nose and throat, and a case could be made for its inclusion in the category of the sternutators (see below, under "Riot-Control Agents"). As a matter of fact, Compton reports that, after witnessing first-hand Germany's use of phenyldichlorarsine and its effects, which include properties as a vomiting agent, England was motivated to investigate diphenyl-chlorarsine (DA) and related arsenicals for its own military use near the end of

World War I. By this time, French military chemists used a mixture of 40 percent diphenylchlorarsine (a solid) and 60 percent phenyldichlorarsine (liquid solvent) as a "mask breaker" called *Sternite*.[25] The former East German military chemist Siegfried Franke reports that German and Italian armies stockpiled mustard mixed with phenyldichlorarsine in World War II.[26]

Blood Agents

Generally speaking, blood agents are poisons that block oxygen utilization or uptake from the blood, effectively causing the body to asphyxiate. Hydrogen cyanide blocks the enzyme critical to aerobic metabolism, and cyanide gas is a major toxic hazard from fires in structures that contain significant amounts of synthetic fabrics, particularly polyacrylonitrile (Orlon), nylon, and polyurethane. It is variously estimated that half of those who perish from "smoke inhalation" do so at least in part because of cyanide poisoning.

The very high volatility of the blood agents makes them less useful as chemical weapons, but their low persistency also has its advantages. Following a concerted battlefield attack with these highly lethal compounds, not much time is needed for the agent to disperse to safe levels, enabling the attacking force to enter the target area. Cyanide may also find a role in terrorism, since hydrogen cyanide gas is easily formed from basic starting materials used in commercial industry.

The blood agents hydrogen cyanide (AC) and cyanogen chloride (CK) have been studied extensively and employed sporadically throughout World War I. At room temperature, hydrogen cyanide (or hydrocyanic acid) exists as a liquid, but it evaporates extremely rapidly. Although it shares toxic properties with hydrogen cyanide (CK is roughly half as toxic), it should be noted that cyanogen chloride also has potent lung irritating effects.

Hydrogen Cyanide: Instrument of the Shoah

Up to this point, most of the chemical agents we've discussed, if they've had wartime applications at all, were employed during World War I. As discussed earlier, there was much less reliance on CW agents by the combatants in World War II. Nonetheless, as we all know, there was a concerted mass campaign of murder during the second war, and the instrument of murder was a chemical agent:

> . . . In the fall of 1941, in compliance with a special secret directive, Russian political officers, commissars and special political functionaries were screened from the prisoner of war camps by the Gestapo and transferred to the nearest concentration camps for liquidation. While on an inspection tour, my deputy SS

> Captain Fritzsch, on his own initiative, used gas to destroy these
> Russian prisoners of war. He crowded the Russians into single
> cells in the basement and, using gas masks, threw Zyklon B into
> the cells, which brought immediate death. . . . Initially, this poi-
> sonous gas—a hydrocyanic acid preparation—was handled under
> extreme precautionary measures only by employees of the Tesch
> & Stabenow Company. . . . During Eichmann's next visit, I
> briefed him on this use of Zyklon B and we decided to use this
> gas for the future mass executions.
>
> —*Rudolf Hess*[27]

Before its adaptation as a tool of genocide, during World War I, France used 4000 tons of hydrogen cyanide against German forces but without achieving much in the way of results. Combinations were also tried using cyanide with arsenic trichloride (30 percent), stannous chloride (15 percent), and chloropicrin (5 percent) to form trichloromethane in the so-called *vincennite* mixture. Even this preparation was highly volatile and therefore not very effective in battle. Under more controlled conditions, gaseous cyanide's massive killing potential was demonstrated decades later in World War II, when hydrogen cyanide was first tested on Russian prisoners of war and then used on millions of imprisoned civilians—men, women, and children. Hydrogen cyanide (or *Prussic acid*) was the active ingredient in the notorious poison Zyklon B, manufactured by IG Farben.

Prone to rapid polymerization and therefore extremely unstable, hydrogen cyanide is today generally understood to be a poor candidate for an effective chemical weapon. Still, in 1939 Germany tested spray equipment for delivering hydrocyanic acid by aircraft at the Münsterlager experimental facility, and was reportedly capable of generating high levels of cyanide that could penetrate protective mask filters. Later, the Soviet Union also attempted the development of cyanide–gas munitions, using an aerial device with 250 kilograms capacity in the late 1930s and early 1940s. In this approach, the rapid evaporation of hydrogen cyanide led to rapid cooling, thus lowering considerably the volatility of the gas. The drawback to this approach was that in order to deliver cyanide gas effectively it would involve dangerous low-altitude flying, and as such aircraft would be quite vulnerable to artillery flak. Further experimentation by the Japanese and American militaries also found problems in the instability of hydrogen cyanide in battlefield conditions.

Unconfirmed reports during the 1980s have suggested that hydrogen cyanide was used by the Syrian government against an uprising in Hama, in an Iraqi military attack on the Kurdish town of Halabja (1988), and in Shahabad, Iran, during the Iran-Iraq War (1980–1988). Little evidence has been published since, however, to confirm these allegations.

UNCLASSIFIED

FIGURE 1

Japanese HCN Frangible Grenade,
Copper Stabilized Type

A Japanese hydrocyanic acid grenade, World War II. An unknown number was thrown at US troops with undetermined effects. (Courtesy of Soldier Biological and Chemical Command, Historical Research and Response Team, Aberdeen Proving Ground, MD.)

Cyanogen Chloride

Cyanogen chloride (CK) is a very volatile compound, but is less a fire or explosive hazard than hydrogen cyanide and therefore logistically speaking less problematic. (Industry has found cyanogen chloride the preferred reactant in processes to make synthetic rubber). Reportedly, France combined hydrocyanic acid with cyanogen chloride in World War I ("manguinite"). The use of cyanogen chloride in this mixture was intended as an irritant to make soldiers remove their masks, exposing themselves to these very toxic gases. Cyanogen chloride was also combined with arsenic trichloride later on in the war. Like hydrocyanic acid, cyanogen chloride tends to spontaneously polymerize and therefore was combined with stabilizers (sodium pyrophosphate) for longer shelf life.

Arsine (Arseniuretted Hydrogen)

Arsine, a highly toxic compound derived from arsenic, falls in the category of a blood agent, but has also been referred to as a nerve poison because of its secondary effects. Much more delayed than cyanide (4–5 hours), arsine poisoning causes the destruction of red blood cells and subsequently tissues of the kidney, liver, and spleen. A poisonous metal known for centuries—and grist for many murder plots in mystery novels—arsenic can be reacted with common chemicals (zinc, sulfuric acid) to form arsine gas. A colorless gas with an unpleasant odor similar to garlic, arsine was studied extensively in World War I. Because of its high volatility and chemical instability, however, it is not considered to have much potential effectiveness as a modern CW agent. As a consequence, not only was the weaponization of arsine largely abandoned in World War I but it (and hydrogen cyanide) also received little attention in later years. As an East German civil defense expert opined in 1956: "Arsine and hydrogen cyanide were given little practical importance even at the start of [World War II] because there was no awareness of the possibilities to produce sufficiently high concentrations of these substances."[28] Arsine is used today for industrial processing of gallium arsenide chips in the semiconductor industry.

Carbon Monoxide

Technically speaking, the common pollutant carbon monoxide could be called a blood agent, as it binds very quickly with the oxygen-carrying iron of hemoglobin. Numerous people die every year in the United States from the carbon monoxide leaked from heating systems or produced by car exhaust. Although carbon monoxide was considered to be a possible weapon in the early twentieth century, in the modern context its high volatility makes it a very unlikely CW agent.

Hydrogen Sulfide ("Sour Gas")

Hydrogen sulfide is more toxic than hydrogen cyanide and was used, albeit marginally, as a CW agent during World War I. Back then it was grouped in a category once termed "paralysants."

Hydrogen sulfide gas is currently encountered as a by-product of decaying organic matter, as a chemical reagent, in volcanic activity, or from the use of sulphuric acid in a variety of commercial processes. In 1997, the FBI suspected that hydrogen sulfide was being considered as a weapon in a plot by four white supremacists determined to rob an armored car. (The would-be robbers planned a diversionary tactic that involved placing bombs on what they believed were storage tanks filled with flammable propane.) As it happened, no hydrogen sulfide was stored at the facility in question, and in any case the plotters seemed to have had no awareness of the toxic nature of the gas. Still, the incident served as a warning that chemicals stored at commercial facilities are vulnerable to terrorists and may pose a risk to people living in the environs.

Nerve Agents (Toxic Organophosphates)

The nerve agents—including tabun (GA), sarin (GB), soman (GD), and VX—produce their toxic effects through both inhalation and contact. So-called "second-generation" CW agents, they are very similar to the organophosphates used in agriculture, and in fact tabun and sarin were discovered in the 1930s by German chemists seeking to make new types of insecticides.[29]

Nerve agents kill by paralyzing the respiratory musculature and can cause death in less than a few minutes. For all intents and purposes, the casualties they produce are immediate. They vary in persistence, with compounds such as sarin creating only a short-term respiratory hazard on the battlefield (on the order of a few hours or more), while more persistent agents such as VX can remain a hazard for many days or even weeks as a ground contaminant.

As mentioned, research on these agents was an outgrowth of work on agricultural compounds. In 1934, Gerhard Schräder of IG Farben was working first with sulfur and fluorine to manufacture cheaper and better insecticides, but on a hunch decided to look at phosphorus, a substance that shared certain properties with sulfur and was also known to have toxic properties. Out of this grew the "German," or G-series, nerve agents, "tabun," "sarin," and "soman," or what have been since respectively labeled GA, GB, and GD in US and NATO code.

During World War II, British scientists were looking to the same configurations of elements for possible use as weapons. The principal British researcher in the field of nerve agents was Bernard Saunders, who often subjected himself to controlled but still risky exposures of these substances to evaluate their

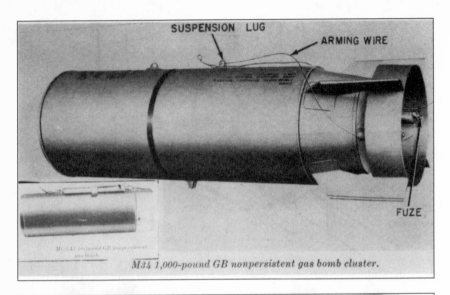

M34 1,000-pound GB nonpersistent gas bomb cluster.

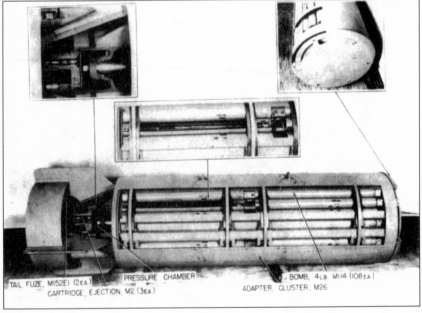

During the Cold War, the M-34 sarin cluster munition was the first major nerve agent weapon for the US military. (Courtesy of Soldier Biological and Chemical Command, Historical Research and Response Team, Aberdeen Proving Ground, MD.)

effects in humans. Saunders discovered that another compound, diisopropyl-flurophosphate[30] (DFP), while not being nearly as toxic as the G-series agents, formed a stable mixture with mustard and could remain liquid at much lower temperatures. Thus, a weapon was conceived that not only included a nerve agent (DFP), but in a kind of synergy could also enhance the performance of a blister agent in winter conditions.

The V-series nerve agents, including VE, VG, VM, and VX (also known as VS), are compounds similar to but more lethal than the G-series. First discovered by a Swedish researcher and characterized in the 1950s by a British chemist (Ranajit Ghosh), these agents tend to be more toxic and more persistent than G-agents. Because V-agents evaporate much more slowly than G agents, they present less of a vapor hazard but a much greater hazard to the skin. Moreover, their high persistence means that they can be used to contaminate (or, in military parlance, "slime") road or ground surfaces in an effort to deny access for days or weeks.

Nerve Agent Proliferation

Nerve agents are the CW agent of choice for most countries and well-funded terrorist organizations pursuing chemical weapons programs because they are among the most lethal and the most suitable for weaponization. Perhaps even more important, they can be produced from the same starting materials—or precursors—used to produce nonmilitary products, including pesticides and flame-retardant compounds. This does not mean, however, that manufacturing nerve agents is a simple matter. Their production, in fact, can involve unpredictable reactions, and in almost all cases highly corrosive compounds are formed as by-products.

A nation or a "sub-state" organization that decides to synthesize its own supplies of a nerve agent in large quantities has to consider the availability of precursors when it decides which agent it will attempt to manufacture. During the 1980s, Iraq discovered that a key precursor for soman (GD), pinacolyl alcohol, was difficult to manufacture with sufficient purity. Faced with this hurdle, the Iraqis abandoned their plans for soman and instead relied on other precursors—including chemicals used to create ordinary rubbing alcohol—to manufacture tabun, sarin, and cyclosarin, as well as VX.

Dynamics of Nerve Agent Poisoning

Nerve agents attack enzymes in the body, and it is this that makes them so deadly. The enzyme that is key to normal autonomic functions as well as muscular contraction (and subsequent relaxation) is acetylcholinesterase (AChE). This enzyme, upon contact with acetylcholine, a key neurotransmitter, normally will cleave off acetic acid to form choline, returning the muscle fiber to a relaxed state. One unit of AChE can, under normal conditions, turn about 300 molecules of acetylcholine into choline within one millisecond. When nerve agents like sarin block the action of AChE, a dangerous trend develops. Fatigued from constant twitching, muscles weaken. In the lungs, bronchoconstriction begins, with associated levels of carbon dioxide rising (acidosis), the lungs begin to compensate by faster breathing. In order to achieve the latter, more acetylcholine is released as the muscles responsible for lung contraction are stimulated. More demands are made of the lungs than the muscles can provide, now weakened by unrelenting stimulation. This vicious circle of nonstop muscle contraction, build up of acetylcholine, and continued stimulation in order to respirate carbon dioxide from the body leads to severe fatigue. Complicating matters further, the increased levels of acetylcholine lead to the constriction of smooth muscle in the respiratory tree, as well as copius amounts of mucosal and salivary excretions, effectively smothering the victim. These in combination with involvement of the central nervous system finally stops respiration completely. Although not necessarily life-threatening, profuse sweating, constricted pupils, urinary incontinence, and skeletal muscle twitching also result from nerve agent intoxication.

In order to gain some appreciation of the lethality of nerve agents, 10–15 milligrams of VX—much less than a drop—in contact with unprotected skin can kill a man of average weight unless he is given medical attention. Still, this route of exposure can take up to several hours before severe symptoms present themselves.

Incapacitants: Psychoactive Chemicals in War

A considerable amount of research and development has been invested in compounds that can incapacitate, disorient, or even paralyze military personnel or civilians. Among others, the following drugs have been investigated for their potential as incapacitating agents:

- BZ
- Lysergic acid diethylamide (LSD)
- Mescaline and its derivatives
- Methaqualone

According to James Compton, an expert on chemical agents and their effects, a proper incapacitating agent must have the following attributes: Its effects would endure over time (hours, days); the agent should not endanger life or cause permanent injury; recovery would not require medical attention or other assistance; and agents must be deliverable, potent, and easy to store. As it happens, very few compounds meet such requirements while remaining suitable for weaponization. After much testing throughout the 1950s and

1960s, the US Army found that other possible candidates, including cocaine, amphetamine derivatives, and nicotine (a very poisonous and psychoactive substance even in small amounts) were not potent enough for aerosol-based weapons. Among the psychoactive incapacitants, the only known agent to have been weaponized for use on the battlefield is BZ (3-quinuclidinyl benzilate). BZ is a member of the belladonna group of compounds (glycolates) that includes atropine, scopolamine, and many others.

Some reports have alleged the use of incapacitating agents by the Soviet Union against the Mujaheddin rebels during its war in Afghanistan. The stories told of a gas that caused the Afghan rebel soldiers to sleep for hours, only to later awake at a Soviet detention facility. Apparently, the agent had a characteristic color and was subsequently named *Blue X*. These charges have not been corroborated. Another report alleged a similar type of "knockout gas" called a *K agent* also used in Afghanistan by Soviet troops. These accounts have not been verified, either, at least in the open literature.

Belladonna, or Glycolate Alkaloids

One of the oldest-known pharmaceuticals (and alternatively, poisons) is derived from belladonna—a poisonous plant of the nightshade family. The earliest use of belladonna to impair enemy is dated to about 200 B.C., when Hannibal's Carthagian army poisoned the wine of rebellious African tribes with mandragora, or mandrake, which contains belladonna-type poisons. Reportedly, in 1672 the Bishop of Münster employed some sort of grenade against his enemies that included a belladonna mixture. (The classic symptom of belladonna intoxication is a widening of the pupils, and this reportedly is the source of the plant's name. Because dilated pupils in women were considered a sign of beauty in ancient Italy and Spain, belladonna plants were often milked for their active ingredients and applied directly to the eyes.)

Atropine is the most widely used belladonna drug today and is often used in ophthalmology for pupil dilation. The myriad effects of atropine and its relatives have included the following classic symptoms, and are easier to remember with the attendant similes: Because belladonna compounds block receptors responsible for glandular secretions, intoxicated individuals are "dry as a bone," and a subsequent lack of perspiration causes the body's temperature to rise, making one "hot as a hare." As a further consequence one becomes "red as a beet" from heat flush. The aforementioned action upon the eye makes normal sight difficult, and therefore one becomes "blind as a bat."

3-Quinuclidinyl Benzilate (BZ)

The belladonna-based incapacitant BZ is an extremely potent compound that attacks the central nervous system. It can also derange a victim's mental faculties, making him "mad as a hatter." Hallucinations are pronounced, often including imaginary or real objects first becoming especially large, then gradually smaller. In BZ experiments conducted on human volunteers, some typical behaviors displayed included plucking at their own clothing ("wool gathering"), as well as imaginary dialogues or fantasies that two intoxicated individuals might engage in, each confirming and playing off the other's hallucinations (*"folie a deux"*). For example, one BZ-treated soldier would offer an imaginary cigarette, which would be graciously refused by the other because "it's the last one in the pack." Another belladonna compound, ditran, has been described as producing more intensive psychoses than those associated with LSD or mescaline (see below).

The history of BZ dates from after World War II. During research into highly toxic nerve agents in the 1950s, the nerve agent antidote atropine attracted a good deal of interest from the military community in England and the United States. In 1952, researchers noted that a similar compound under study had even stronger pharmacological effects than atropine.[31] This substance was code-named "BZ" (also called *agent buzz*)[32] and was weaponized by the United States in the 1960s as an incapacitating agent. Technically speaking, BZ was considered a central nervous system depressant and a hallucinogen and was designed primarily to be thermally disseminated in an aerosol fog.

Between 1962 and 1964, BZ was manufactured for military use at the US Pine Bluff Arsenal in Arkansas. But because it was both highly toxic and unpredictable, BZ quickly fell out of favor in the US military, even before the implementation of the Chemical Weapons Convention in 1997. Most, if not all, remaining stocks were destroyed in the 1980s.

BZ did not, however, disappear from the battlefield. In a 1998 report, Human Rights Watch alleged that it was used against Bosnian refugees by the Yugoslav People's Army (JNA) during 1995. Iraq reportedly developed an agent similar (if not the same) to BZ called *Agent-15*.

Ergot and Lysergic Acid Diethylamide (LSD)

1951, France: The calls from stricken new patients kept flooding in. By Monday, August 14, the town's hospital was swamped. Seventy homes had also been turned into emergency wards. That first night the first victim died in agonizing convulsions. Raving patients were held in their beds or escaped from their homes, mad, frantic, to run in the streets. The terror grew as the news broke that a demented eleven-year old boy had tried to strangle his own mother. The mood of the people and the atmosphere of the place began to resemble that of a plague-swept town of the Middle Ages. . . . Finally, the chief toxicologist of Marseilles sent

his report to Pont-St.-Esprit and the anxious nation. The bread
contained twenty alkaloid poisons, three of them virulent, and all
came from the same source. The poisons could be found in fungus growth that changed normal kernels of rye to purplish cock-
spurs called ergot.

—Famine on the Wind [33]

Several major diseases have been associated with the human consumption of
moldy rye, including spasmodic and gangrenous ergotism. The latter, as its
name suggests, is characterized by gangrene, in this case in the limbs, and particularly the fingers and toes. Once prevalent in Europe, ergotism is a classical
illness caused by moldy grain. A fungus (*Claviceps purpurea*) growing on the rye
builds up levels of ergot poison in the plant as a means to survive harsh winter
conditions, and while it affects many grains, rye is particularly susceptible.
Hundreds of episodes of mass poisonings involving ergot have occurred
throughout history, including a reported 40,000 deaths in France during the
year 994. In the early twentieth century, ergot was commonly used as an aid
for expectant mothers experiencing difficult births. In these cases it was called,
among other names, Ergotin or Bonjean's Ergotin, and was considered then to
be somewhat unreliable.

In 1938, Swiss researchers at Sandoz, using ergot as a base, synthesized an
analogue named lysergic acid diethylamide (LSD or LSD-25). What they discovered about its effects was not published until 1947. Grouped in the category of indole derivatives, LSD is a white, crystalline product that has
extremely powerful hallucinogenic and behavior-modifying properties.
Lysergic acid amide compounds have also been isolated from the psychoactive
Mexican plant *ololiuqui*, which is used in shamanistic rites. These psychoactive
compounds make up 0.01 to 0.05 percent by weight of its seeds. Since at least
the 1950s, ergot and its derivatives, especially LSD-25, have been researched
for their potential as both human intelligence gathering tools (in the CIA's
MKULTRA program in the 1950s) as well as sabotage poisons. LSD is
extremely potent, many times more so than other compounds (for example,
mescaline) that have been suggested for sabotage or diversionary use. Data in
humans are somewhat contradictory, but it would appear that only 20 micrograms of LSD are needed to cause symptoms that would otherwise require 15
milligrams of mescaline (a near thousand-fold difference).[34] Because battlefield application would require large amounts of this highly potent drug, it is
likely that many would die from its exposure, contradicting the original purpose—to only incapacitate personnel. Since the drug would not reliably
intoxicate targets though the skin, LSD would require delivery in the form of
an aerosol or contaminating fragmentary weapons. However, it could have a
role in terms of harassing or sabotaging enemy logistical areas. Delivered in a

food or beverage to a small number of people, however, it could have an extremely disruptive effect.

Mescaline and Its Derivatives (Phenyl Ethylamines)

Forming a group of adrenaline (norepinephrine) acting compounds, mescaline can be derived from the peyote cactus (*Lophophora williamsii*) found in the southwestern United States and northern Mexico. (The buttons of the cactus plant are made into a tea or ingested to produce hallucinatory episodes.) By mimicking adrenaline and over-stimulating the central nervous system, mescaline and related compounds produce hallucinations that are roughly similar from one individual to the next: brilliantly-colored lights and geometric shapes, but not the derangement and delirium associated with BZ intoxication. Close relatives include amphetamine analogues ("speed") such as MMDA or ecstasy, a drug that has found use (and abuse) among recreational users within the last decade. Trimethoxyphenylaminopropane (TMA) is another compound that has similar properties to mescaline, producing hallucinations in smaller doses and psychoses that last about 6 hours. Both TMA and MMDA are synthetically produced.

Despite its potency as a psychoactive compound, mescaline has similar attributes to LSD with regard to its physical and chemical properties, therefore making it difficult to weaponize. Furthermore, mescaline is generally more toxic and would be difficult to employ without resulting in many deaths, defeating its original purpose as an incapacitating agent.

Methaqualone

No longer considered much of a military threat, methaqualone—a central nervous system depressant with the street name *Quaalude*—has been widely abused as a recreational drug. It was produced in great quantity by the South African Wouter Basson in the 1980s and 1990s. (It is not clear exactly what the intention was in the South African case, although it could have simply been a matter of illicit narcotics trade disguised as a chemical weapons program.) Related to the barbiturates, a category of drug that continues to have medicinal use as a widely prescribed pharmaceutical, methaqualone was discontinued as a prescribed drug in the west by 1984.

Harassing or Riot-Control Agents (RCAs)

During the last three decades, debate has continued over the ethics of using riot-control agents (such as tear gas and vomiting agents) and herbicides (for example, Agent Orange) in wartime operations. In a 1969 resolution of the United Nations General Assembly, 80 nations signed a declaration to proscribe "any chemical agents of warfare," including herbicides and riot-control agents. Herbicidal agents, while not restricted under the 1993 Chemical Weapons Convention, are prohibited when used as a form of warfare (see below). Riot control agents (RCAs) are now considered CW agents under the CWC if they are used "as a method of warfare," although their use is permitted for domestic law enforcement. Otherwise, the RCAs described below are considered under the general rubric of chemical weaponry, and their use in war is prohibited.

Most often, RCAs are used to combat public disturbances or as a non-lethal means to disperse unruly crowds. One incident in May 1951 during the Korean War involved North Korean POWs who had become agitated, and American troops resorted to tear gas in order to regain control. (This action was quickly seized upon by the communist governments of North Korea and China, propagandizing this action as an atrocity involving CW agents when they brought the case to the United Nations.) Some US Army operations during the Vietnam War also employed RCAs to protect landing zones or to flush out Viet Cong guerrillas from tunnels. It was, in fact, the controversial use of tear gas in Vietnam that helped start the sequence of events leading to the vociferous protests against herbicides in Southeast Asia.[35] CS and CN were authorized for use in controlling Iraqi POWs during Operation Desert Storm (1991), but this turned out to be a largely unnecessary measure.

Tear gases or RCAs can irritate the eyes and the mucosal tissues of the nose and mouth, causing extreme discomfort in humans (some animals, such as horses, seem to be resistant to at least some RCAs). Some of these non-lethal compounds are also called lacrimators,[36] the name suggesting the shedding of tears and immediate and extremely painful effects upon the eyes. (In the United States, a similar group of nose and throat irritants were sometimes referred to as sternutators, or "sneezing" agents). While we have classified some here as RCAs, the categorization is somewhat arbitrary, as some (such as CN) originally were conceived as veritable CW agents. And even though they are not strictly speaking lethal unless ingested in high doses, RCAs can cause discomfort and panic, and in enclosed spaces, panic can express itself in lethal ways.

Having largely replaced CN many years ago, CS (named after its inventors, Corson and Stoughton) is the most commonly used riot-control agent today. Another agent, CR, is more potent in terms of its effects (causing coughing and watery eyes), but less toxic than CS; not enough information is yet available to warrant its routine use for crowd control.

With the exception of capsaicin (originally proposed as a potential harassing agent in World War I) or pepper spray, tear gases are often a solid at room temperature and are dispersed in an aerosol (a suspension of fine particles in the air), in a liquid solvent (for example, Mace™), or as vapor generated using high heat from a canister. While insoluble in water, CS can be readily neutralized in simple, water-based solutions (unlike CN), and the effects normally do not last beyond 20 minutes of leaving the contaminated area. CS offers advantages in its high potency at low levels, but is conversely less toxic than other riot-control agents.

Because the effects of RCAs mimic (although nonlethally) those of standard CW agents, the RCAs listed here appear in the same categories used for more conventional CW agents: lacrimators (eye irritants), sternutators, and vomiting agents (D-series of arsenicals).

Lacrimators (Eye Irritants)

As early as 1912, French constables found that ethyl bromoacetate—an extremely irritating substance—was useful in controlling violent mobs, including increasingly boisterous labor groups. It was not much of a stretch, then, to take this application of a riot-control agent to the use of lacrimators in World War I by French soldiers, some of whom had earlier served in the Paris constabulary. Perhaps taking this cue, Germany and other participants in World War I developed agents of their own that caused extreme eye irritation, utilizing halogenated organic compounds (i.e., adding chlorine, bromine, and iodine to their chemical structures). In fact, the first substantial (if not successful) CW attack by Germany was not chlorine but xylyl bromide (*T-Stoff*), technically an eye irritant of the lacrimator class. But even predating the German shell attack in January 1915 were previous uses of eye irritants by the French military. All told, some 6000 tons of lacrimators were used in World War I.

As one goes down the list, it becomes apparent that a number of organic chemicals were systematically halogenated, tested for their effects on humans, and subsequently applied to the battlefield to harass the enemy. Few of these compounds are especially toxic, save for phenylcarbylamine chloride, which has potent lung irritating effects. Chloracetophenone, now commonly known as Mace™, was never used in World War I as it was only developed after the Armistice. But all shared the characteristic of "producing almost instantaneous physiological effects (in less than one minute) in the form of a muscular reaction of the eyelids, closing of the eyes, and a glandular reaction from the lacrimatory glands, producing a copious flow of tears."[37] Because only chloropicrin (see above) and chloracetophenone are seen much nowadays, we only treat some of the World War I lacrimators in brief:

Ethylbromacetate. In 1858, Perkin and Duppa combined bromacetic acid and ethanol to form ethylbromacetate. This compound earlier found service by French constables in 1912 to incapacitate criminals and expedite their capture. Officially the first "combat gas" employed in World War I, ethylbromoacetate was delivered in the form of rifle grenades (26 millimeters) by France in

November 1914. Some German commentators alleged France had prepared the use of this compound long before hostilities broke out, but this was denied by the French. (Germany also tested dianisidine chlorosulfate, a dye base derivative, in October 1914 but with little success.) More toxic than chlorine gas, the effects of ethylbromoacetate on the eyes are so potent that few would probably stay in the affected area for long.

Chloracetone. In November 1914, chloracetone replaced ethylbromoacetate mentioned above in French grenades, both for tossing in the form of greandes and firing from rifles. Though this proved to be a strong lacrimator, gas masks developed early on in World War I seriously degraded its effects on the battlefield.

Xylyl bromide ("T-Stoff"). Early in World War I, when there was no shortage of bromine in Germany, its military scientists experimented with a number of organic compounds, xylyl bromide being the first true CW agent deployed by the *Wehrmacht* in January 1915. Among other problems encountered by this novel approach, Germany discovered that xylyl bromide corroded most metals, requiring lead canisters to hold the liquid for artillery shells. Less toxic than ethylbromacetate, xylyl bromide is much more potent as a lacrimator, detectable in concentrations as low as 0.0018 mg (or about 2 micrograms) per liter. Though the effectiveness of xylyl bromide especially during the winter was questionable, its use in artillery shells opened the door for more advanced chemical weaponry to follow.

Benzyl bromide (also coded "T-Stoff"). Called *Cyclite* by the French, this close derivative to xylyl bromide was formed by reacting toluene with bromine. It was, however, somewhat weaker than xylyl bromide, and it too quickly found disfavor as a war gas.

Bromacetone. Coded *B-Stoff* by Germany, *Martonite* by the French, and *BA* by the Allies, bromacetone found most widespread use (1000 tons or more) in World War I for its intense lacrimatory effects. Its high volatility, while a drawback in other respects, meant that high concentrations could be created simply by lobbing many shells at a given target, allowing the vapors to rise and spread. Largely supplanted by bromobenzyl cyanide in 1918, France and Great Britain continued to make mixtures of bromacetone and chloracetone through the duration of the war.

Brommethylethyl ketone. During much of World War I, England faced a severe acetone shortage, a shortcoming that was overcome by the timely inventiveness of Chaim Weizmann. Germany was also feeling the pinch and looked for other organic substances besides acetone to create new lacrimators or other toxic warfare agents. Using methylethyl ketone and brominating it, Germany called this new substance *Bn-Stoff*, while France called theirs *Homomartonite*, owing to its similarity to bromacetone (see above). Germany first used brommethylethyl ketone in July 1915 (following the German chlorine gas attack in

April). For all of its economic advantages, brommethylethyl ketone never made a serious impact on the battlefield, with bromacetone still being the most effective lacrimator used.

Iodoacetone ("Bartonite"). First used by the French in artillery shells in August 1915, iodine was reacted with chloracetone to form iodoacetone. (Iodide was substituted due to short supplies of bromine.) Benzyl iodide (see below) was found to be superior a few months later, and iodoacetone was quickly abandoned by France.

Ethyliodoacetate ("SK"). Generally speaking, iodine has usually been more costly to acquire than chlorine or bromine. In World War I, however, England found its supplies of bromine running low, while having a temporary glut of iodine (imported from South America). Thus the invention of ethyliodoac-etate, first used by the British at the battle of Loos in September 1915. When delivered by 4.2-inch howitzers and 4-inch Stokes mortars, alcohol was mixed with the compound to increase its volatility on the battlefield.

Benzyl iodide ("Fraissite"). Faced with its own shortage of acetone, and with supplies of bromine also running low, France, like England, combined benzyl chloride with potassium iodide to form benzyl iodide, a very potent lacrima-tor but with low volatility. For that reason as a weapon it was delivered in a half-and-half mixture of benzyl chloride and benzyl iodide.

Acrolein ("Papite"). Also driven by economic exigencies, France turned to glycerine for production of a lacrimator in 1916. This compound required none of the costly halogens such as iodine or bromine for its mass production. Problems soon arose, however, when it came to weaponizing acrolein as a CW agent. As the name suggests, acrylic acid can be readily formed by exposure of acrolein to air, and subsequently polymerizes into an inert gel. French chemists tried adding 5 percent amyl nitrate to stabilize the compound, but this created yet another problem: spontaneous formation of acrolein gum. Though a potent lacrimator and relatively toxic, technical problems prevented it from becoming an effective CW agent.

Bromobenzyl cyanide ("Camite," US/NATO: "CA"). Riener synthesized bro-mobenzyl cyanide by halogenating (for example, adding bromine) phenyl cyanide, with industrial production commencing in 1914. First used by the French in July 1918, the US military also adopted this extremely potent lacrimator (several more times stronger than bromacetone), manufacturing it in quantity by fall 1918. But like other lacrimators that preceded it, bro-mobenzyl cyanide corroded metal, was prone to rapid decomposition, and had low resistance to heat. Thus, delivering bromobenzyl cyanide in artillery shells required special containers, as well as a smaller charge as a fuse, otherwise little agent would remain following shell detonation. In his classic study of chemical weaponry in World War I, Augustin Prentiss saw few roles for bromobenzyl cyanide as a CW agent in future conflicts.[38] However, one may see this compound referred to from time to time as a riot control agent, but here too its threat is usually downplayed.

CN (chloracetophenone). The German Carl Graebe discovered the intense, lacrimatory effects of chloracetophenone in 1869.[39] While all sides during World War I had access to the formula and its accompanying scientific literature, mass-scale production was difficult and only achieved by war's end. Being a solid substance in its pure form, when devising effective weapons chloracetophenone also presents more engineering challenges. Coded CN by the United States, chloracetophenone is among the most potent lacrimators known, while having relatively low toxicity in mammals. During its military's suppression of indigenous rebels on Taiwan in 1930, Japan may have been the first to not only use chloracetophenone as a means of warfare, but also holds the dubious distinction as being the first to use CW in modern Asia.[40]

Chloracetophenone can be delivered in three basic ways: detonated as a crystalline powder for dispersion, thermally generated with burning gunpowder, or dissolved in an organic solvent. The more effective means of delivery for warfare would be heat-generated smoke, such as the classic mixture of one part CN and three parts gunpowder, delivered in a shell or bomb. Marketed as a personal defense weapon, Mace™ (i.e., *M*ethylchloroform chloro *ace*tone) is essentially CN dissolved in liquid for spraying against an assailant. Pepper spray (using the capsaicin derived from chili peppers) is now replacing CN for individual use and by law enforcement. For larger-scale use in riot control, CS (see below, under "Sternutators") has shown to be more effective while having a lower toxicity profile than CN.

Sternutators

During military operations in Korea and against Cypriot rioters in the 1950s, anecdotal information has it that the British introduced CS after finding its earlier CN and chloropicrin mixture ineffective. Named after its creators Corson and Stoughton who synthesized ortho-chlorobenzylidene malononitrile in 1928, CS is now the RCA, or tear gas, of choice for the United States and many other countries, replacing the use of CN by the US military in 1959.

The following is an excerpt from a study that described the effects of CS agent on a group of human volunteers:

> Typically, men leave the exposure with tears, nasal secretions, and saliva pouring out, and towels rather than handkerchiefs are needed to cope with the fluids. In 5–15 minutes, the irritation ceases.[41]

CS can be delivered in two basic forms: CS1 and CS2 consist of microencapsulated particles that are aerosolized in particles ranging 3–10 microns in diameter. The other technique employs a thermal grenade (for example, the M7) that generates CS fumes. The only major difference between CS1 and CS2 is the type of desiccant used to keep the CS particles from agglutinating.

Some of the dangers of CS are especially to be found when used in enclosed spaces. The tragic aftermath of US law enforcement's siege in Waco, Texas, of the Branch Dravidian compound, was probably in part due to CS particulates that had earlier been poured directly into the building during the standoff. Like the fine dusts that have been the source of grain elevator explosions, with enough oxygen interspaced between particulates a sudden fire can result. While reports have suggested that cynanide can be liberated from CS upon its entry into the body, it does not appear to be a significant factor in CS toxicity, at least from its delivery as an aerosol.[42]

Vomiting Agents

The effectiveness of RCAs in war was made especially apparent in 1917, when arsenic-based compounds were added to munitions in order to harass the enemy. At this time, Germany employed diphenylchlorarsine in a mixture of diphosgene and phosgene gas. A year later, diphenylcyanoarsine or in a mixture of diphenylchlorarsine was found to be even more potent. These were termed "vomiting agents" due to their effects in humans. Even more so than other RCAs, vomiting agents can also cause death in certain instances when used in confined areas. By common agreement among its members, none of these compounds were ever permitted to be used by NATO countries.

DM (Adamsite) and related arsenicals. Vomiting agents such as Adamsite (DM, diphenylaminearsine) cause extreme pain to the eyes, nose, and throat, and these symptoms can be followed by vomiting and bowel constriction after a few minutes exposure. These form a part of the arsenical group of substances mentioned earlier in reference to Lewisite (see above), including DA (diphenylchlorarsine) and DC (diphenylcyanoarsine). Until it was supplanted by CN tear gas, Adamsite had been the more commonly produced vomiting agent in the United States and elsewhere.

Banned RCAs

The United States renounces, as a matter of national policy, first use of herbicides in war except use, under regulations applicable to their domestic use, for control of vegetation within US bases and installations or around their immediate defensive perimeters and first use of riot control agents in war except in defensive military modes to save lives.

—*United States Air Force, February 1998*[43]

The CWC now prohibits the use of RCAs against combatants in wartime. The United States, however, has reserved the right to use RCAs in certain combat situations and under the following circumstances:[44]

- Counter-terrorist and hostage-rescue operations
- Noncombatant rescue operations outside of internal or international armed conflicts
- Military operations within an area of ongoing armed conflict when the state is not a party to the conflict (such as the United States and Somalia, Bosnia and Rwanda)
- Consensual peacetime operations when the receiving state has authorized the use of force (including Peacekeeping operations under Chapter VII authority of the United Nations Security Council)

The reasons for such exceptions made by the US government is due in large part to the experience of the Vietnam War, when CS gas was used as a means to suppress enemy fire during the extraction of downed American pilots in the jungle. RCAs are also widely viewed as a humane method of dispersing civilians who may be intermixed with combatants during paramilitary operations.

Herbicides

Although not considered by all to be chemical weapons, herbicides (or defoliants) are not proscribed under the CWC unless they are used as a "method of warfare." We include a brief listing here of herbicides used in civilian applications as well as war:

Herbicide	Compound
Paraquat	Bipyridylium
Agent White	Picloram, 2,4-D
Agent Orange	2,4,5-T and 2,4-D
Agent Blue	Dimethyl arsenic acid

In both World War II and later during the Malaysian "Emergency" of the 1950s, herbicides were developed for dual roles: to deny the enemy cover from foliage and to destroy crops being utilized by enemy forces. Agent Orange and other herbicides were used in similar fashion throughout the Vietnam War. Much of the work in defoliants actually started under the aegis of the US biological warfare (BW) program, for although these chemical compounds and their development were still considered part of the overall CW arsenal, the BW organization was chosen to carry out herbicide research "as a matter of scientific economy."[45]

Obscurant Smokes

Many historical references—and even some modern allegations of chemical warfare—are no doubt instances of non-lethal smokes of various kinds, employed to confuse the enemy. For example, during China's ancient period it is said that the rebel Chi You created a fog to confuse his southern enemies. This smoke caused such havoc that were it not for Emperor Huang Di's "directional chariot"—a legendary vehicle that could navigate in darkness— the Northern barbarians might very well have won that day.[46] Using a "screening smoke" in 1701, King Charles XII of Sweden effectively shrouded his forces moving across the Duna river against Russia.[47] Although the use of smokes in war is sometimes treated in CW-related texts, phosphorus oxide and other smokes are not very toxic and thus not considered chemical weapons as such. However, the presence of smoke on the battlefield can sometimes be construed as *prima facie* evidence of a chemical weapon attack when actually no CW agent was used.

Used widely in World War I, white phosphorus was discovered to be a useful compound to generate large amounts of smoke to camouflage military operations. White phosphorus, however, can be quite toxic and has been largely replaced by titanium tetrachloride for generating obscurant smokes.

Napalm

Napalm is often mentioned in various places along with chemical weapons, but this is due primarily from its development by chemical warfare services around the world. Put another way, because it required special chemical expertise, napalm has been organizationally handled or at least controlled in some fashion by CW-related departments. But it is primarily an incendiary fuel and, technically, not a chemical weapon. Its development, however, is of some historical interest and in some ways coincides with modern chemical weaponry.

Used some 1500 years ago, Greek Fire was certainly an ancient precursor to modern incendiaries. The recent history of napalm goes back to a "jellied" mixture of gasoline and rubber developed at Harvard University in October 1941. The M-47, an incendiary bomb produced in 1941–2, utilized gasoline jelly with ignition supplied by white phosphorus surrounding the TNT-tetryl burster. As World War II erupted into the Pacific campaigns, rubber supplies dropped to nil, and other thickening agents as possible substitutes were then investigated by Standard Oil, Arthur D. Little, and the du Pont Company. A combination of aluminum soaps of naphthenic and coconut fatty acids (NAP of *na*phthenic and PALM of *palm* oil for Napalm) made a thickener that was extruded with gasoline. About 80,000,000 pounds of Napalm were made in

World War II for incendiary bombs and flame-throwers. Later, isobutyl methacrylate polymers were employed as thickeners, which later were also employed to make true and veritably toxic CW agents more persistent.

Malodorous Concoctions and Masking Agents

Early ordnance manuals from the nineteenth century European militaries include stink ball recipes, consisting of sulfur, parts of horse hooves, and other items that would produce a bilious, overpowering stench. Smoke generated from such mixtures would have been somewhat toxic depending on the amount of exposure. In World War I, foul-smelling substances were investigated to mask the odor of toxic compounds, such as mustard agent, or to simply cause psychological stress for the opposing forces. After the war, a Colonel R. F. Bacon of the American Chemical Warfare Service had this to say on the subject:

> The gas-camouflage is of particular interest. It has been found
> that malodorous compounds (butyl mercaptan, dimethyl tricar-
> bonate, etc.), are useful to mask the presence of other "gases" or
> to force the enemy to wear respirators when no other "gases" are
> present.[48]

Everyone has probably noticed the strong smell of skunk while driving along a rural highway. The noticeable odor of a skunk's essence, as it were, is present even at very low concentrations and is scientifically known as the group of mercaptans, or combinations therof (consisting largely of trans-2-butene-1-thiol and 3-Methyl-1-butanethiol, depending on the species). A similar compound to that of skunk is n-butyl mercaptan.[49] It is a straight chain, 4-carbon molecule with a sulfur and hydrogen attached at the end. (The "n" stands for normal, or the indication that the carbons are in a straight chain conformation.) N-butyl mercaptan is the compound mentioned above in Victor Lefebure's classic book on chemical weaponry, *The Riddle of the Rhine*. (Similar compounds have also been implicated in the "skunkiness" of beer; one reported mechanism is that of light filtering through clear or green glass beer bottles. Green glass, in particular, allows a frequency of 400–520 nanometers, cleaving off a molecule contained in hops, the main flavoring ingredient in beer, that subsequently reacts with hydrogen sulfide. The result is a slight odor of skunk.[50])

In the US chemical weapons program, some very foul smelling compounds were also investigated, including derivatives of skatole, which, as the name suggests, is a foul-smelling substance akin to feces or rotting offal. An official US medical textbook on CBW reports, however, that the relative ease of protecting against such odors (for example, wearing masks) and the probability that a highly motivated enemy would not be appreciably deterred by aversive odors alone caused this line of investigation (which was never very popular with the research team) to be abandoned.[51]

Although not necessarily toxic in and of themselves, obnoxious odors can harass or camouflage other toxic compounds when used in military or possibly contexts involving terrorism. For example, there have been over a hundred recent cases involving "noxious chemical vandalism" on abortion clinics in the United States. In all of these, butyric acid was used. Butyric acid is the active compound that produces the foul odor in rancid butter and can be irritating to the skin and eyes in high concentrations, but generally is not very toxic. But it does require a lot of effort to get rid of the lingering odor.

CHAPTER 5

Chemical Warfare: A Brief History

We generally think of chemical warfare as a modern phenomenon, but it has its precursors in ancient and medieval warfare, and especially siege warfare. Assaults on castles and walled cities could drag on for months or even years, so it is not surprising that a military commander would look for an innovative way to end a stalemate. Incendiaries and toxic or greasy smokes were common tools in both attacking and defending besieged castles.

Perhaps the first recorded use of poison gas occurred in the wars between Athens and Sparta (431–404 B.C.). The Greek historian Thucydides reported that flaming pitch (tar) and sulfur mixtures were employed during the Peloponnesian War, these being ancient examples of fiery siege weapons and smoke-generating concoctions.[1] (If inhaled in sufficient concentrations, sulfur-based smokes can be quite toxic). Thucydides also made reference to Spartan armies deploying the toxic metal arsenic in vapor clouds.[2]

In the fourth century B.C., the Greek military scientist Aeneias Tacticus recommended the use of smoke to deter siege miners. Around the same period, in pitched battles among vying Chinese states, noxious smokes were employed to defeat underground sappers. According to the historical classic *Mo Zi*,[3] firewood and hemp were bound together, set alight, and lowered by a chain into to the enemy mines. Later variations suggested a combination of grass, reeds, and firewood could be dropped "on the hole outside the city wall where they are mining to smoke and burn them. The enemy will immediately die."[4] The Chinese *Gunpowder Epic* (*Wu Jing Zong Yao*) recounts the addition of arsenic in making incendiary bombs, a practice that may have been common before A.D. 1000.

In a classic engagement in 189 B.C., the Roman army laid siege to the Greek city of Ambracia. Consul Marcus Fulvius, after having surveyed the city, first attempted to break down the walls of the Greek stronghold with battering rams. When this proved to be unsuccessful, miners were called upon to break the wall's foundation from underground. The Ambraciots, however, had their own plans, and placed brass plates along counter tunnels near sections of the wall that were under Roman attack. By observing the intensity of the metal plate vibrations, they were able to dig their own counter-tunnels close to the Roman miners. Then, a large clay vessel filled with burning coals and bird feathers was connected to a pipe, which was in turn shoved through the earth to penetrate the Roman tunnel. By attaching a bellows to the smoke pot, the Greeks were able to blow noxious fumes into the Roman sappers' tunnels, driving them away

FROM GREEK FIRE TO THE *FLAMMENWERFER* . . .

Fire and smoke have a long history as a tool of warfare. Petroleum is perhaps the oldest known incendiary used in large-scale combat. Assyrian bas-reliefs dating to the ninth century B.C. show what is believed to be liquid petroleum being used as a fire-assault weapon. Aeneias Tacticus refined the combination of tar, sulfur, and pine resin for use as an incendiary against warships in 360 B.C.

Naphtha, a mixture of hydrocarbons that bears some resemblance to gasoline, was distilled from crude oil by the Arabs as early as the sixth century. The combustible properties of naphtha and its utility as a weapon was probably first brought to Byzantine Rome's attention by the inventor and architect Callinicus in about A.D. 668, when he traveled to Constantinople and taught the Romans the secret technology of *Greek Fire*. Because they are liquid and volatile mixtures of hydrocarbons, naphtha and other petroleum distillates are inefficient as weapons, since it is very difficult to control exactly when and where in relation to a target they will combust. To compensate for these shortcomings, Greek Fire made use of wax and oil of balm, which were mixed in with the fuel to add thickness and sticking ability. In a later technical improvement on Greek Fire, quicklime (calcium oxide) was added to create a delayed incendiary. When moistened, the subsequent mixture of fuel and quicklime could reach a temperature of over 150 degrees Celsius, spontaneously igniting the composition after it had adhered to the target (say, a naval vessel).

As early as the eighth century A.D., Arab soldiers were using naphtha as a weapon, and by the time of the Second Crusade (A.D. 1147–1149), Arab military units called *naffatun*, clad in specially designed fire-protective clothing, were organized to make incendiary attacks. They ignited containers filled with thickened fire oil, wax, sulfur, and plant fibers (tow) to defend against the besieging European forces. (It is worth noting that gunpowder was thus far completely absent in these military engagements.)

The ancient and medieval worlds of Europe and the Middle East had discovered the military advantages of their version of napalm long before they knew of gunpowder. Between the period of Muslim expansion and the first

Crusades, contacts were made between the Islamic world and East Asia. A tribute of 86 bottles containing *"meng huo you"*[5] (literally "fierce fire oil") was made by the King of Champa (in modern day Cambodia), Sri Indravarman II, to the court of the Late Zhou dynasty in Kaifeng, Henan province. The historian Joseph Needham[6] makes a good argument that this "fierce fire oil," sent to the Chinese by way of a diplomat named Abu'l Hassan, was in fact naphtha, the same basic compound used in Greek Fire.

And this is where gunpowder (or an early version of it) enters the picture. The use of the naphtha mixtures in a military setting required a potent and reliable ignition source. Edward Vedder, the author of *Medical Aspects of Chemical Warfare*, suggested that this "Liquid fire was never used successfully by anyone but the Byzantine emperors, and attempts by others to utilize it have uniformly failed."[7] However, we find what could be the very first military use of gunpowder inextricably linked to the use of Greek Fire by Chinese naval warships in the Song dynasty (in the tenth century A.D.). At least by A.D. 919, the Chinese were using gunpowder—with smaller amounts of nitrate in proportion to sulfur and carbon—in order to contrive a specially designed wick. By means of a system of compression pumps, naphtha was squirted out of nozzles and allowed to brush past the slow-burning match that had been impregnated with gunpowder. A Chinese historian describes how the powder-fired flame-thrower worked in this excerpt from the *Gunpowder Epic*:

> ... [B]efore use the tank is filled with rather more than three
> caddies of the oil with a spoon through a filter; at the same time
> gunpowder (*huo yao*[8]) is placed in the ignition-chamber at the
> head. When the fire is to be started one applies a heated branding
> iron (to the ignition-chamber), and the piston rod is forced fully
> into the cylinder—then the man at the back is ordered to draw
> the piston-rod fully backward and work it (back and forth) as
> vigorously as possible. ... If the enemy comes to attack a city,
> these weapons are placed on great ramparts, or else in outworks,
> so that large numbers of assailants cannot get through.[9]

Like Roman and Greek combatants before them, the Chinese flame-throwers probably found their most effective use during naval engagements, such as a massive campaign on the Yangtze River in A.D. 975. The venerable Chinese scholar Shi Xubai writes about this famous battle in his historical treatise *Talks at Fisherman's Rock*:

> ... Zhu Lingpin[10] (admiral of the Southern Tang) was attacked by
> the Sung emperor's forces in strength. Zhu was in command of a
> large warship. ... Zhu Lingpin hardly knew what to do. So he

quickly projected petrol from flame-throwers to destroy the
enemy. The Sung forces could not have withstood this, but all of a
sudden a north wind sprang up, and swept the smoke and flames
over the sky towards his own ships and men. As many as 150,000
soldiers and sailors were caught in this and overwhelmed, Zhu
Lingpin, being overcome with grief, flung himself into the flames
and died.[11]

Chinese use of gunpowder in the eleventh century A.D. included "thunder-clap bombs"[12] and "thunder-crash bombs,"[13] the latter munitions using larger proportions of potassium nitrate in the gunpowder to create bombs of higher shock energy (or "brisance"). This all happened before the English friar and scientist Roger Bacon is known to have experimented with gunpowder in 1267. Nonetheless, one could at least say with confidence that Roger Bacon was the first in Europe to codify the properties and preparation of black powder. Although his knowledge of black powder seemed to have concerned itself solely with pyrotechnics, Bacon was in fact concerned that its explosive properties might fall into the wrong hands, and he chose to write its formula in a cryptic cipher. Or so the story goes. We do know that he came into possession of firecrackers, and that his contemporaries were known to have visited the Mongol courts of the Khan, so it is almost a certainty that the explosives were of Chinese origin.

Another story is that the fourteenth-century German monk and alchemist Berthold Schwartz was the inventor of modern firearms. According to this version of events, it was in Freiburg, Germany, about a century after Roger Bacon's forays into the alchemy of explosives, that "the black Monk" Berthold married gunpowder with a workable firearm. Berthold's ultimate fate is uncertain. He may have either accidentally blown himself up—the most probable outcome, if he had in fact existed at all—or even been executed because of his infernal discoveries. In fact, very little documentation of his life exists at all:

> ... history may have taken no interest in his doings because guns
> were said to be execrable inventions and their employment
> (except against the unbelievers) was decried as destructive of
> manly valor and unworthy of an honorable warrior. Berthold was
> reputed to have compounded powder with Satan's blessing, and
> the clergy preached that as a co-worker of the Evil One he was a
> renegade to his profession and his name should be forgotten. ... [14]

While purists would not consider the early use of *Greek Fire* and gunpowder-based incendiaries as true chemical warfare (CW), these early flame- and smoke-producing techniques have direct (and not-so-direct) connections with the modern use of toxic substances on the battlefield.

The Nineteenth Century

No doubt in time chemistry will be used to lessen the
suffering of combatants.

—*Sir Lyon Playfair, 1854*[15]

In the nineteenth century, Admiral Lord Dundonald of the British Navy concocted a plan involving the use of sulfur fumes during the Crimean War (1853–56). At about the same time, Sir Lyon Playfair proposed the use of cyanide in artillery shells to be fired on Sebastopol. The British War Office as well as the British Admiralty were adamantly opposed to these ideas, rejecting them as being "against the rules of warfare." In rebuttal, Playfair wrote:

> ...There was no sense in this objection. It is considered a legitimate mode of warfare to fill shells with molten metal which scatters among the enemy, and produces the most frightful modes of death. Why a poisonous vapor which would kill men without suffering is to be considered illegitimate warfare is incomprehensible....[16]

When the eminent chemist Michael Faraday (1791–1867)[17] was asked by the British government to advise how to employ toxic chemicals against Russia in the Crimea, Faraday replied that such a plan was possible, but that he also thought it barbaric and would have no part in such a venture.

The American Civil War (1860–1865) also saw the use of an early form of chemical warfare, in the use of countermining smoke. Following the detonation of an 8000-pound black powder bomb placed underneath Confederate Army troops (killing over 300 troops), rebel soldiers at Petersberg dug countermines to prevent further surprises. Colonel William W. Blackford, serving as an engineer with cavalry troops led by Jeb Stuart, had his men dig holes of 4 inches in diameter that would penetrate the Union mines. The plan was to have Confederate soldiers on lookout for any Union subterranean activity, and if any were detected, Blackford wrote, the rebels would counterattack:

> ...the guards on duty were provided with cartridges of combustibles, the smoke from which would suffocate a man. These they were to run into the holes and fire by a fuse, closing their end of the hole tightly, and then, summoning the guard, they were to dig into and take possession of the opposing mine as rapidly as possible, giving another dose of suffocating smoke from time to time to keep the enemy out of his workings until they could dig into them.[18]

Wyndham Miles, a historian of chemical weaponry, surmised that this plan involved the use of powder with a high sulfur content to generate high concentrations of sulfur dioxide. He also concludes by noting that Blackford's scheme was the only case of actual CW in the entire Civil War. That may be true, but not for lack of other plans. In an 1862 letter sent to the US War Department by a certain John Doughty of New York City, there is a description of a type of chlorine shell that would "render conflicts more decisive in their results." In many ways, Doughty's scheme was prescient, describing the essentials of chemical warfare using chlorine gas:

> ... If the shell should explode over the heads of the enemy, the gas would, by its great specific gravity, rapidly fall to the ground: the men could not dodge it, and their first intimation of its presence would be by its inhalation, which would most effectually disqualify every man for service that was within the circle of its influence; rendering the disarming and capturing of them as certain as though both their legs were broken. . . .[19]

Another idea came from a geologist of some renown, Forrest Shephard, whose scheme (never carried out) was presented in a letter to President Abraham Lincoln. Shephard recommended that clouds of hydrochloric acid be generated to harass Confederate troops:

> ... [B]y mingling strong sulphuric acid with strong hydrochloric, or muriatic acid on a broad surface like a shovel or shallow pan, a dense white cloud is at once formed, and being slightly heavier than the atmosphere, rests upon the ground and is high enough to conceal the operator behind it. This may easily be continued by additional sprinkling of the two acids and a light breeze will waft it onward. When the cloud strikes a man it sets him to coughing, sneezing, etc., but does not kill him, while it would effectually prevent him from firing a gun, or if he should fire, to aim at his object.[20]

In Europe, too, thought was given to the ways in which chemistry could be harnessed to make weapons more effective. In 1813, during the "War of Liberation" from Napoleon, a pharmacist suggested to the Prussian General Friedrich Wilhelm von Bülow that hydrocyanic acid (HCN) could be applied to bayonet blades by means of small brushes. As Jean-Pascal Zanders, a noted CW authority, points out, it is difficult to conceive how such a venture could have been practicable. HCN is quite volatile, does not remain very long on any object exposed to the elements, and would just as likely poison friendly troops in any case.

And during the Franco-Prussian War (1870), a scheme to apply cyanide on bayonets was also recommended, only this time to the Emperor Napoleon III, but the technique was never carried out for use in battle. The nineteenth century saw, then, the sporadic use of what one could call chemical weapons, as well as a number of proposals for using toxic chemicals that were mostly never carried out. While there was certainly a reluctance by military powers to resort to the use of poisons for war, certainly not all were so disinclined, and one needs to think about why it took so long for the advent of truly modern CW to finally appear in World War I. In other words, what made the Great War of 1914 so different from, say, the Crimean War of 1853–1856? The answer has a lot to do with the trend toward industrialization and the development of a chemical industry that moved science from the laboratory to the factory.

The Dawn of Organic Chemistry

Before getting into the development of modern CW agents, we need to take a brief look at classical organic chemistry as it existed in the early 1900s. In 1915—the year that ushered in modern CW—organic chemistry as a scientific discipline was about 50 years old. Unlike today, when just about every conceivable substance used in the chemical industry can be ordered from a catalog, in the 1800s the basic starting points were oil and coal-tar. What you needed to make, you had to synthesize from either or both of these hydrocarbons.

From coal-tar, which is derived from bituminous coal, one can synthesize products as different as pesticides, plastics, explosives, and food flavorings. When separated into its constituents, coal turns out to be a mixture of a huge array of useful compounds. While it had been known for centuries that flammable vapors could be derived from coal—and these gases had already been utilized to provide street lighting in many modern cities by the mid-nineteenth century—it was only much later that many of the valuable substances within coal-tar were identified with precision. Distilling tar from coal yields, among other compounds, benzene, toluene, phenol, and xylene, and these in turn can be used to create a multitude of products. Toluene, for example, is critical for the manufacture of explosives, namely trinitrotoluene (TNT), while phenol is used in the manufacture of many drugs as well as pesticides and insecticides. And these and many of the other derivatives of coal-tar, as it turned out, were useful in the synthesis of dyes. The industrial revolution that began in the eighteenth century had created a massive surge in textile manufacture in the nineteenth, and along with it an unprecedented demand for clothing dyes. Up until this time, inks and dyes for textiles were almost exclusively derived from plants. For hundreds of years, civilizations had used natural indigo in textiles, art, and printing. The natural dye is extracted from, among other species, the leaves of the plant *Indigofera sumatrana*. But in the early nineteenth century it was found through chemical analysis that natural indigo dye, upon being distilled, broke into two main compounds, indole and aniline. The German chemist Adolf von Baeyer (1835–1917) determined the chemical structures of indigo and its constituents—a remarkable achievement, especially for the time—and he was subsequently awarded the 1905 Nobel Prize in Chemistry for his work. His discovery of the molecular structure of indigo and his research on many other organic compounds did much to develop the German chemical industry, particularly the dye-manufacturing and drug-manufacturing industries—both of which relied completely on the synthesis of organic compounds. Moreover, the research done in the laboratories of each industry spurred on and fed research efforts in the other, within Germany and throughout Europe. For example, in England in 1856, William Henry Perkin, a then 18-year-old prodigy in chemistry, discovered the purple aniline dye mauve (*Mauveine*) accidentally, while he was working in a lab attempting to synthesize the anti-malarial drug quinine. And the German pharmaceutical giant Bayer AG was originally founded (by Friedrich Bayer and Johann Friedrich Weskott, in 1863) not as a drug company but a dye-manufacturing and -marketing firm. Even BASF, which today is still one of the world's dominant chemical firms, started out as a manufacturer of synthetic dyes.

In the late nineteenth and early twentieth century, Germany came to dominate most markets dependent on organic chemistry. Although France and England had earlier made considerable contributions to the science, Germany quickly surpassed them in this nascent field through the concerted and coordinated effort of its universities, industries, and government. This was made embarrassingly evident in 1914, on the eve of World War I, when Britain was forced to purchase dyes from German suppliers to produce uniforms for the Royal Navy.

Beginning in 1880, BASF undertook an 18-year research project to synthesize indigo on a commercial scale, and spent about 18 million marks before it was able to sell this new man-made dye on the marketplace. When Germany finally succeeded in devising a viable synthetic process for indigo, Fritz Haber, who was later to become the father of modern CW, came to a rather startling conclusion: the natural dyes were headed toward obsolescence. He was right. Whereas in 1871, over 1.5 million acres of Indian land were devoted to indigo farming, by 1914 only 150,000 acres were utilized for this purpose. Once a viable commercial process for producing indigo dye had been achieved, the problem was how to acquire sufficient quantities of the starting materials and intermediate compounds necessary to set up a mass-production operation. The starting material, naphtha, or naphthalene—a coal-tar derivative—did not present much of a problem, since coal was abundantly available in Upper Silesia (now part of Poland). The difficult part was the industrial manufacture of reactants, especially ethyl chlorhydrin, which in turn required production methods for liquefied chlorine and ethylene. (Though more well-known for his work in physical chemistry, Fritz Haber also conducted research in the chemistry of hydrocarbons, anticipating modern techniques of "cracking" petroleum derivatives to produce, among other things, ethylene.)

Meanwhile, in 1886, at his Göttingen laboratory, Victor Meyer was making use of ethyl chlorohydrin[21] by reacting it with sodium sulfide, from which he produced an oily but otherwise unremarkable compound called thiodiglycol.[22] Using his product (a common dye intermediate used today), Meyer then ran a reaction that attached two chlorine atoms. What happened next was completely unexpected. There was nothing in these compounds or in their apparent nature that indicated a vicious poison would result. Nonetheless, an assistant of Meyer's who was in charge of the synthesis was nearly killed by the toxic product of this experiment.

At first, Meyer was not convinced that a toxic compound had been formed, wondering whether or not his assistant had just been overreacting— or, less charitably, the apprentice had some sort of mental problem. To learn more, Meyer sent the product to a medical college, where his substance was applied to rabbits. The animals, when exposed to the vapors, developed conjunctivitis (irritation of the inner eyelid and outer eye membranes) and then died. Burns on rabbit skin produced by the chemical resembled those that developed on Meyer's hapless assistant. Meyer subsequently wrote:

> . . . The intended work with this chloride was not continued—on
> account of the extremely poisonous qualities of the compound. It
> is very striking that this apparently harmless substance which is
> only slightly volatile, is almost insoluble in water, and has a very
> slight odor as well as a perfectly neutral reaction, should exert a
> specific toxic effect. Its chemical constitution would never lead
> one to expect its aggressive properties. . . .[23]

There the matter rested until 1912, when Hans T. Clarke, a chemist who had also studied in Berlin, decided to duplicate Meyer's synthesis. A flask of the stuff broke, and Clarke suffered a severe chemical burn on his leg and required two months of hospitalization. Clarke had himself modified the synthesis route slightly, replacing phosphorus trichloride with hydrochloric acid, and it was the latter method that Germany would use to make mustard on a massive scale during World War I. After the war, Clarke told the US Chemical Warfare Service that it was his own laboratory mishap—promptly and dutifully reported to the Chemical Society—that may have inspired Germany to develop mustard as a CW agent.

By 1916, when Fritz Haber became chief of its chemical warfare service, Germany had clearly known of mustard's high toxicity and capacity for use as a weapon. Germany could not have produced significant quantities of mustard without ethyl chlorhydrin, a chemical they were able to divert from its commercial use in synthetic dye production. Haber knew that the Allies would be starting from a near-zero capability in terms of responding with chemical weapons of their own, and Germany's lead in chemicals would give it a great advantage. Even so, Germany waited for supplies of mustard to accumulate in order to use the agent decisively. Haber advised his government that if Germany wanted to use mustard, it should do so with the aim of winning the war quickly. He was not one to underestimate the technical prowess of the Allied nations and knew that Germany had at most one year before the United States, France, and the United Kingdom would be able to respond in kind.

When Germany first employed mustard agent at Ypres in 1917, samples from the front were examined by French and British scientists. Figuring out what the substance was didn't pose too much of a problem—the trick was how to produce it themselves. The chlorhydrin used by Germany to produce mustard on a large scale was not available on a large scale to Great Britain and France, and by the time the Allies had devised mustard production facilities of their own, the war was nearly over. (Approximately 75 percent of the mustard that was eventually used by the Allies was produced at one facility in France, but only with great cost and effort. France finally was able to use mustard offensively in June 1918, and Great Britain in August that same year, mere months before the Armistice.)

From the *Flammenwerfer* to the Livens Projector: The Buildup to War

There have always been expressions of moral disapproval to greet the invention of new weapons technologies, and especially at the introduction of flame-projecting weapons. In World War I, Germany's invention of the *Flammenwerfer*—a very large version of the flame-thrower—was widely held to be a damnable creation, on a par with poison gas, at least according to prevailing opinion at the League of Nations. Winston Churchill, reflecting upon World War I, included flame-throwers along with gas as exemplars of barbarity:

> . . . Bombs from the air were cast down indiscriminately. Poison
> gas in many forms stifled or seared the soldiers. Liquid fire was
> projected upon their bodies. Men fell from the air in flames, or
> were smothered often slowly in the dark recesses of the sea. The
> fighting strength of armies was limited only by the manhood of

their countries. Europe and large parts of Asia and Africa became
a vast battlefield on which after years of struggle not armies but
nations broke and ran. When all was over, Torture and
Cannibalism were the only two expedients that the civilized
Scientific and Christian States had been able to deny
themselves. . . .[24]

Churchill himself, of course, had championed the aggressive use of CW
against Germany—his wife had once approvingly referred to him as a
"Mustard Gas fiend."[25]

The entry of the German *Flammenwerfer* into World War I, in July 1915, was
met with consternation and terror, although one witness reported that the set-
back suffered by British troops was due "more to the surprise and temporary
confusion caused by the burning liquid than the actual damage inflicted."[26]
The introduction of this large flame-throwing weapon resulted in a concerted
attempt by the British to respond in kind, and to an unintended consequence.
The engineering of an Allied version led directly to the Livens Projector,
named after the colorful and brilliant inventor Major William Howard Livens
of the British Army. The Livens projectile is roughly 2 feet in height and about
7.5 inches in diameter, with a combined weight (including gas or explosive) of
60 pounds, half of that consisting of chemical agent. The projector could
launch large volleys of chemical shells in rapid succession, and it was capable
of hurling combustible oil over 90 yards, making it the most efficient means of
deploying gas shells during the war.

At the start of 1916, the so-called Special Sections of the British Army were
reorganized into Z Company, and Livens was put in charge of four gas (CW)
sections. Like his father, Livens had pursued a career in civil engineering, and
his background prepared him well for the technical features of weaponry.

Early in the war, Livens supervised trials of a flame-thrower invented by
Captain F. C. Vincent from the British Ministry of Munitions. The prototype
was cumbersome and far too dangerous because it employed oxygenated fuel.
(It exploded during tests under Livens's supervision.) In late 1916, Livens lob-
bied for the adoption of a new design, and he and his father together came up
with several. One was the size of a large butter churn; it never saw actual service. The other, a decidedly non-portable version, did make it to the Front. It
required a hole dug several yards deep, roughly a yard wide, and 15 yards in
length. Reservoirs of incendiary fuel sat atop a long tube, with a running
length of a small tube that supplied pressure. Resembling a massive pop-up
irrigation sprinkler system, this tube ran underground to a vertical fixture
called a monitor. The monitor had a piston-type nozzle that upon receiving
pressurized oil pushed itself through the thin layer of ground and spewed fuel
toward the enemy. Once through the surface, the surging oil was lit from a

flame placed beneath the stream. Ten seconds of firing created a huge flame, but used approximately one ton of a crude oil-distillate mixture. Although moderately successful in both killing enemy soldiers and suppressing enemy fire, the effort involved in transporting the weapon to the front, not to mention the effort to install it, created logistical nightmares. This British *Flammenwerfer* was thus abandoned, having only ten firings throughout World War I. Still, Livens was to take what he had learned from the flame-thrower and put it to use on behalf of the British Army.

The Livens Projector

I connect up one lead
Ha Ha Ha Ha
I connect up the other one
Ho Ho Ho Ho!
What care I if the zero pass
So long as I can give the Bosche a dose of gas
Then it's over the top and camouflage
No [Royal Engineer] could be bolder
But when they shell, we run like hell
And dump the old exploder

> — *"Pooping Off," an anonymous "poem"*
> *written by a British soldier in World War I*[27]

One day during an attack at the Somme, Livens's Z Company came across dug-in German soldiers. After grenades showed little effect on the Germans, Livens made a Molotov cocktail out of two 5-gallon oil cans and tossed them into the shelter.

> The effect was so good that [Harry] Strange thought it would be
> a better plan to throw the oil over to the Bosch [i.e., the
> Germans] in the original packages in preference to the labourious
> method of discharging it from the elaborate flame-thrower.[28]

Having all but given up on their *Flammenwerfer*, Livens and his group turned their attention to the idea of delivering incendiary oil in containers. Projectiles filled with oil were fired using modular sack charges, a method by which one could adjust the range by adding or subtracting set quantities of explosive propellant. In comparison to the very complicated flame-projecting contraption, this device was extremely simple. For all intents and purposes, it was essentially a kind of mortar, and it was its very simplicity that turned it into an extremely effective weapon.

The Stokes mortar (designed in 1915) was used to launch chemical as well as other ordnance. (Courtesy of Soldier Biological and Chemical Command, Historical Research and Response Team, Aberdeen Proving Ground, MD.)

While its first use was for firing incendiaries into enemy trenches, it soon became clear that the Livens Projector, as this weapon was now known, could also be used to deliver poison gas. It was a significant improvement over releasing gas from static cylinders, and the rather large projectiles could carry significant quantities of CW agent with decent accuracy. When filled with a CW agent such as phosgene, it rivaled the Stokes mortar, a tube artillery piece that up until 1917 was the chief delivery device for chemical agent shells. A captured German document revealed the following:

> . . . The enemy has combined in this new process the advantages
> of gas clouds and gas shells. The density is equal to that of gas
> clouds, and the surprise effect of shell fire is also obtained. . . .
> Our losses have been serious up to now, as he has succeeded, in
> the majority of cases, in surprising us, and masks have often been
> put on too late.[29]

The projector/mortar was set into the earth at a 45-degree angle to maximize the range, which generally was more than 1 kilometer. Because most of its features were hidden by earth, the weapon was relatively easy to conceal, and provided some cover for the soldiers firing it. It also proved to be remarkably reliable, and by 1917 was in use on all fronts. In fact, the Livens Projector was so successful that the German Army quickly followed with its own version, a *Gas Projector*, later that year.

Chemistry That Changed the World

> Dr. Weizmann, I was one of the mightiest men in Germany. I was
> more than a great army commander, more than a captain of
> industry. I was the founder of industries; my work was essential
> for the economic and military expansion of Germany. All doors
> were open to me. But the position which I occupied then, glam-
> orous as it may have seemed, is as nothing compared to yours.
>
> —*Fritz Haber to Chaim Weizmann, 1933*[30]

By 1916, German military and political leaders began to face a matter of grave concern: they desperately needed to break the stalemate because they lacked a sufficient supply of conventional munitions, particularly nitrogen-based explosives. Germany's war effort depended on being able to stockpile these weapons, and the stockpiles were in turn dependendent on supplies of one elemental component, nitrogen. Up until the early 1900s, the world depended in large measure on nitrates imported from Chile, the main component of nitrogen-based fertilizers and explosives. In 1913, Fritz Haber discovered a method of making ammonia from atmospheric nitrogen, known today as the

Haber-Bosch process.[31] Synthesizing ammonia from the nitrogen that is abundant in the atmosphere would mean that Germans could secure large amounts of it relatively cheaply[32] and locally, critical issues in the years of war that were soon to follow.

At the same time, England was suffering from a shortage of the solvent acetone, another important ingredient for munitions (cordite). Winston Churchill asked a distinguished scientist, the Russian-born and German-trained chemist Chaim Weizmann, a British subject, for help. "Dr. Weizmann," Churchill is reported to have said, "we need thirty thousand tons of acetone. Can you make it?"[33] The desperate situation was solved when Weizmann, later to become the first President of Israel, led a team that utilized bacteria that naturally produced the acetone. Plants to "breed" the solvent were constructed in both Great Britain and the United States, including an acetone production facility at Terre Haute, Indiana.

When asked by the British government what kind of compensation he desired for his invaluable assistance, Weizmann, a Zionist, asked for help in establishing a Jewish homeland. The very creation of Israel (by means of the Balfour Declaration, in 1917) was in part a gesture of gratitude to the chemist on the part of Great Britain.[34]

Germany's war effort, too, may have been saved from defeat by a chemist. At the start of World War I, both Fritz Haber and a rival, Walther Nernst, served as scientific consultants for the German military. Earlier, during civilian life, Nernst had blocked Haber's appointment as professor of physical chemistry at the University of Leipzig. Nernst also told his students that he was really the one responsible for the Haber process of fixating nitrogen. Polite in public, there was no love lost between these two chemistry professors, and Haber would soon have professional revenge.

The first serious attempt to use chemicals in combat began with a German artillery barrage against the French, using (ortho-) dianisidine chlorosulphonate, or *Niespulver*—sneezing powder,[35] and it was Nernst who had suggested its use. In October 1914, *Niespulver* was filled into 3000 shrapnel artillery shells, and these were fired at French positions in Neuve Chapelle. Although dianisidine is somewhat toxic and can irritate the mucous membranes, it has only limited ability to create casualties on the battlefield, and the attack failed miserably.

As Nernst's sneezing powder debacle unfolded, Haber was trying to prove his worth to the German Ordnance Department. The challenge that had been put forward to Haber was to formulate a gasoline mixture that could withstand freezing temperatures. By using a combination of xylene and naphtha, Haber was able to produce an effective anti-freeze for winter fuel, as well as to gain the trust of those in charge of the Ordnance Department. He was assigned next to work on a program investigating the use of chlorine as a gas weapon, and by February of 1915 had produced experiments that looked promising. By then, Haber was the *de facto* successor to Nernst.

WORLD WAR I

World War I, the Great War, the first "global" war in history, was in its first few months almost unimaginably destructive and deadly. By the end of 1914, a mere four months after it began, the conflict had produced over 600,000 fatalities and was averaging 150,000 new dead each month. It is difficult even to conceive of such carnage. By means of comparison, France alone had lost more men (306,000) in these four murderous months than did all South Vietnamese armed forces during ten years of the Vietnam War (275,000), or the US armed forces lost in World War II (295,000). Initially, when the great armies had assembled in late July 1914, military leaders anticipated that combat operations would be highly mobile, and that the war would run its course rapidly. During the first few months of the war this was in fact largely the case. Fritz Haber himself noted that no one in Germany, himself included, had foreseen the stalemated trench warfare that would soon predominate.[36]

While Great Britain was slow to understand the value of the digging of trenches, even after receiving hard lessons during the Boer War (1899–1902) Germany saw the advantages of earthen fortifications very quickly. By September 14, 1914, the German army's "fortification and defense" of the Aisne river in northern France had set the stage. By the end of 1914, an unbroken line of trench works would run the western Front from Switzerland to the North Sea—a distance of about 475 miles—and for the next three years the main belligerents were locked in a murderous stalemate. Lines were soon dug in, forming broken segments of trenches, parapets, and shell holes that ran miles in length. Opposing trenches ranged from hundreds to only about 25 yards apart from each other. By 1915, the trenches would extend for almost 1300 miles. Machine guns—and improved accuracy with other firearms—made movement extremely dangerous and charges against enemy lines suicidal. Said one historian of World War I:

> . . . To show one's head over the parapet was to commit death. It should be borne in mind that artillery and rifle fire never ceased by day or night—it was only a matter of degree. . . .[37]

For Fritz Haber, the man who practically invented the concept of modern CW, the trenches and the standoff they helped create called for the use of chemical agents. Wrote Haber,

... Life in the trench—subject to direct hit or cave-in—is a terrific strain on human nerves, but the experience of the war has
taught us that the strain becomes tolerable because sensitivity is
deadened, as it is deadened against any continuous stimuli on the
human organism. . . . Exactly the reverse is true of the means of
chemical warfare. [Its] essential characteristic is the multifold and
varying physiological effect on man, and the sensations they produce in him. Any change in the impressions felt by nose and
mouth affects the psychic equilibrium through the unknown
character of the effect, and is a new strain on the power of moral
resistance of the soldier, at a time when his entire psychic strength
should be devoted undividedly to his mission in combat. . . .[38]

The Chlorine Attack at Ypres

... "The French have broken," we exclaimed. We hardly believed
our words. The story they told we could not believe; we put it
down to their terror-stricken imaginings—a greenish-gray cloud
had swept down upon them, turning yellow as it traveled over the
country, blasting everything it touched, shriveling up the vegetation. No human courage could face such a peril. Then there staggered into our midst French soldiers, blinded, coughing, chests
heaving, faces an ugly purple color—lips speechless with agony,
and behind them, in the gas-choked trenches, we learned that
they had left hundreds of dead and dying compadres. The impossible was only too true. It was the most fiendish, wicked thing I
have ever seen. . . .

—Rev. O. S. Watkins, Ypres, Beligium, April 1915[39]

In the midst of massive carnage, many in the German scientific community
were convinced that chemical agents could be used to break the stalemate.
Driven by a desire to see a quicker conclusion to the war, Haber organized the
first major chlorine assault against the Allies. At his direction, on April 22,
1915, the *Wehrmacht* released a barrage of chlorine gas against Allied forces at
Ypres. The first chemical assaults were highly organized and massive operations. Imagine the logistical nightmare: 90-pound cylinders, carried by hand
over treacherous terrain, often at night to conceal their placement. During the
initial chlorine attacks, 5730 cylinders (approximately half of the current supply of chlorine in Germany at the time) were buried at the front lines with
pipes leading out into no-man's-land. Stretched across a front of 6 kilometers,
the cylinders released their contents (German soldiers had to do this by hand)
when the wind direction was appropriate, and clouds of gas drifted over the
enemy positions.

The French Army (in this case, mostly made up of Algerian troops) and Canadian soldiers suffered thousands of casualties at this first attack. Although probably not the 5000 or so usually claimed in the histories, the effects were no doubt impressive, with at least 800 dead and 2000 wounded.

Because the attack at Ypres was for all concerned a new form of battle, even experimental in nature, the Germans were not prepared. Their reserves of manpower were low, and they therefore were not prepared to take advantage of their success. Furthermore, by not pressing its hand at this first attack, Germany lost the element of surprise. In later attacks, both sides took primitive—but somewhat effective—protection measures, making chlorine less and less effective throughout the remaining years of World War I.

The problems posed by chlorine were not confined to the French forces who were under this chemical attack. The agent is a liquid when stored in pressurized metal cylinders, but when the valve on the cylinder is opened, the sequence of events can be disastrous for attacker and defender alike. With virtually no experience to go on, the first attacks involved German soldiers simply opening up the valves, allowing the prevailing wind to carry the gas toward the enemy. With hundreds of cylinders lined up, most of the soldiers were able to open their tanks at the same time and quickly fall back. But many did not. One of the things that happens when liquids evaporate is a loss of heat, including in surrounding materials. As the chlorine evaporated, the metal cylinders cooled down so quickly that the valves froze, becoming nearly impossible to turn, and soldiers frantically trying to get their valves open soon found themselves in the midst of a deadly cloud of gas.

In later attacks, modifications were made to the tanks in order to solve some of these problems, and chlorine was mixed with other irritants in the hope that enemy soldiers could be forced to take off their gas masks. Eventually, chlorine, as an armament, quickly fell out of favor as a weapon with both sides in the conflict. But it had done its damage.

Mustard Enters the War

. . . During the night of October 13–14th [1918] the British opened an attack with gas on the front south of Ypres. They used the yellow gas whose effect was unknown to us, at least from personal experience. . . . About midnight a number of us were put out of action, some for ever. Towards morning I also began to feel pain. It increased with every quarter of an hour, and about seven o'clock my eyes were scorching as I staggered back and delivered the last dispatch I was destined to carry in this war. A few hours later my eyes were like glowing coals, and all was darkness around me.

—*Adolf Hitler,* Mein Kampf, *1924*[40]

Chlorine gas was first used by Germany against Allied forces in April 1915. This photograph shows a French attack using chlorine, which is blown by the wind toward the German trenches. (Courtesy of Soldier Biological and Chemical Command, Historical Research and Response Team, Aberdeen Proving Ground, MD.)

First tested in the summer of 1916 and employed at Ypres in July 1917, mustard was a significant departure from volatile liquids and gas. In low concentrations, mustard can damage the eyes and lungs, and it forms blisters upon contact with the skin. With its latent and insidious action, mustard caused the greatest numbers of wounded throughout World War I.[41] Just one month after its introduction on the battlefield by Germany, British casualties from mustard exposure were almost equal to *all* gas casualties from the previous years.[42] Both German and Allied armies discovered that filling CW agents into artillery shells and special gas-bombs made the weapons less weather-dependent, easier to target, and therefore more lethal. Artillery shells filled with a persistent agent like mustard were to be used with great effect throughout the remaining years of World War I. In one engagement alone, over 50,000 mustard shells were fired, some of these shells containing nearly three gallons of the agent. Mustard more than earned its infamous title as "king" of the chemical warfare agents.

Weapons Used and Abandoned

All told, approximately 124,000 tons of chemical agent munitions were used in World War I, most of these being delivered in some 65 million artillery shells. At least 20 percent of the chemical munitions were duds, and about 13 million rounds of chemical shells were left behind. Many of these shells are still scattered throughout former battlegrounds in Europe and on the ocean floor.

When World War I ended in 1918, over 16 million acres of France were cordoned off due to the danger of unexploded ordnance. Today, more than 80 years after the conflict, many chemical bombs and shells still remain scattered in the former "No Man's Land" in France, requiring special engineers—*démineurs*—to dig up and destroy countless munitions posing hazards to local inhabitants and farmers. Most of this ordnance contains high explosive, but some may also have remnants of CW agents such as mustard.

Even with a vintage of 40 years (or more), mustard contents can remain highly toxic. In the 1950s an accidental burst from an old mustard shell from World War I killed two children, while severely injuring several others. Even as recently as 1990, after handling a mustard shell left over from the war at Verdun, an elderly Frenchman suffered serious burns on his hands and arms. As a tribute to those who fought and died in the first major cauldron of chemicals in battle, a museum was built in 1998 near Poelkapelle, Belgium, that displays trench warfare re-creations and artifacts from the battlefields of Ypres.

With protective gear for both men and horses, this was a typical World War I field artillery unit of the Allied forces. (Courtesy of Soldier Biological and Chemical Command, Historical Research and Response Team, Aberdeen Proving Ground, MD.)

THE AFTERMATH: PERSPECTIVES
ON CHEMICAL WARFARE

Indeed, when we attempt to interpret atrocity in terms of available casualty statistics, we find that gas is slightly less atrocious than other weapons.[43]

The judgement of future generations on the use of gas may well be influenced by the pathetic appeal of Sargent's picture of the first "Mustard Gas" casualties at Ypres, but it must not be forgotten that in looking at the picture that 75 per cent of the blinded men he drew were fit for duty within three months, and that had their limbs and nerves been shattered by the effects of high explosive, their fate would have been infinitely worse.

—General H. Hartley, March 1919[44]

We have already described how the combatants first imagined that World War I would be a quick war, and a confident Germany thought she had enough conventional weapons to bring it to a rapid conclusion. The grueling war that dragged on as millions were killed brought a sense of helplessness to both sides. As historian John Keegan tells us, it was the battle in November 1914 at Ypres, Belgium, that brought to the opposing forces the realization that the war would be one of "attrition, mass death and of receding hope of victory."[45] Breaking through the enemy's defenses of earth, wood, and concrete trenches—and then being able to capitalize on or even preserve any such gains—were to be the great challenges of the entire war.

In such a milieu of grand slaughter, the combination of military strategy with Germany's advanced chemical industry made modern CW almost inevitable. In World War I, chief of staff General Erich von Falkenhayn thought the idea of using chemical weapons "unchivalrous," but he overcame his own objections in the hope of a quick resolution of the war.[46] Even Fritz Haber, the father of modern CW, thought the idea horrific, as did his wife, who committed suicide upon learning her husband would return to the front to direct another chemical salvo. But Fritz Haber also sincerely believed that poison gas could end the war and alleviate suffering on both sides.

But suffer they did. In the aftermath of the war, thousands of veterans were left horrifyingly mangled not by chemical but conventional weapons. The scale of human disfigurement by bullets, shrapnel, and shells required an immense post-war effort to manufacture prosthetics not only for limbs, but to replace noses and entire lower jaws. As a result of this gruesome wartime legacy, many World War I contemporaries—especially those who participated in founding the first Allied chemical warfare services—saw the vociferous opposition to chemical warfare as ill-informed. Apologists for CW maintained

These examples of chemical agent delivery were typical of the 1920s and 1930s in the US arsenal. From left to right: the 75-mm mustard shell, the 4.2-in. white phosphorus shell, the M1 30-lb mustard bomb, the Mk II 155-mm mustard shell, the Livens phosgene projectile (CG = phosgene), and the Mk I portable chemical cylinder. (Courtesy of Soldier Biological and Chemical Command, Historical Research and Response Team, Aberdeen Proving Ground, MD.)

that chemical weapons were at least no more inhumane than conventional weapons, and that the real horror was caused by bombs and rifle slugs rather than gas.

It is certainly true that (contrary to popular belief) long-term illnesses due to chemical exposure during World War I were actually minor. In 1924, during hearings before the US Senate concerning the Veterans' Bureau, Albert P. Francine spoke to this issue, testifying that "the permanent effect of gas is, I believe, more serious on the morale and on the heart."[47] The causes for such conditions among veterans, including chronic bronchitis, were subsequently determined to have been complications from concurrent infections (such as tuberculosis) or smoking. According to Curt Wachtel, former advisor to early twentieth century German and Russian military CW programs, the relatively low mortality figures from World War I chemical weapons spoke volumes:

> There can be no doubt that the comparison between these fig-
> ures: mortality through gas, 1.73–4.2%; mortality through all
> other weapons, 24–30%, proves that gas is the most humane
> weapon ever used so far.[48]

Shortly after the war, German, American, French, and British apologists for the use of chemicals in war often compared CW to the horrors caused by more "acceptable" weaponry. The following is from a report by the Surgeon-General of the United States in 1920:

> Gas is twelve times as humane as bullets and high explosives. That
> is to say, if a man gets gassed on the battlefield he has twelve
> times as many chances to get well as if he is struck by bullets or
> high explosives.[49]

These are grim comparisons, and the words of Victor Lefebre quoted at the beginning of this section may perhaps say it best: all one can say is that one type of weapon is slightly less atrocious than another. By the end of World War I, it is fair to say, Europe was exhausted—politically, economically, even emotionally. There were fitful and largely unsuccessful efforts to make a lasting peace, and to control the types of weapons that nations were permitted to develop (see Chapter 6), but peace was not to last, even for a few years.

Tukhachevsky and the War Against the Peasants

Members of the White Guard bands, partisans, bandits, surrender! Otherwise, you will be mercilessly exterminated.

—*General Mikhail Mikolaevich Tukhachevsky, 1921*[50]

The complete story of the brutality that characterized the first decades of Soviet rule still remains to be discovered, but here we will try to shed some light on at least one facet of the history: the role chemical weapons played during Lenin's suppression of the Kulaks (farmers).

In 1921, although having largely defeated the White Army, the Soviet Red Army was facing a rebellion in Tambov Province, in central Russia. So severe was resistance among these peasants, led by a guerilla-styled fighter named Alexander Stepanovich Antonov, that a General (and later Marshall), Mikhail Mikolaevich Tukhachevsky, was called upon to take action. In May 1921, over 100,000 troops, including Chinese and Hungarian volunteer forces, were mustered for operations in Tambov. Said a contemporary Bolshevik who later recalled the campaign against Tambov, "It was decided to conduct all operations in a cruel manner so that the very nature of the actions [taken] would command respect."[51]

For the "pacification" of the peasants, the running theme for Tukhachevsky and the secret police was "no mercy." The ferocity of Red Army reprisals for the peasants' resistance were biblical in scope. In an order signed by Tukhachevsky (No. 171), among other commands were listed the need to "shoot on sight any citizens who refuse to give their names," and that "wherever arms are found, execute immediately the eldest son in the family."[52] Right after having issued Order No. 171, Tukhachevsky anticipated the need for poison gas. "The remnants of the defeated rebel gangs and a few isolated bandits," Tuckhachevsky wrote,

> are still hiding in the forests . . . the forests where the bandits are
> hiding are to be cleared by the use of poison gas. This must be
> carefully calculated, so that the layer of gas penetrates the forests
> and kills everyone hiding there. The artillery inspector is to pro-
> vide the necessary amounts of gas immediately, and find staff
> qualified to carry out this sort of operation.[53]

Tukhachevsky then warned the Tambov peasants and their rebellious kin that the Red Army would, among other punitive actions, "smoke the bandits out of the forests" with the use of "asphyxiating gas."[54] What sort of agents were in mind for such a campaign? It is possible that combinations used by Russia during World War I may have been contemplated, these carrying a number of lung irritants such as phosgene, chloropicrin, tin tetrachloride and chloracetone. However, it has yet to be revealed whether or not the Red Army actually used gas against the Tambov rebels. Other alleged uses of chemical weapons against the Soviet people, such as the claim that CW agents were disseminated by aircraft in the republics of Central Asia in 1930, also await further documentation.

The Wushe (Paran)[55] Incident:
The First Use of Chemical Weapons in Asia?

Defeated by the Japanese in 1895, per terms of the Treaty of Shimonoseki, the Manchu rulers were forced to cede the provincial island of Taiwan to Japan. During Japan's colonization of Taiwan, brutal pacification campaigns were waged against local indigenous groups, particularly during the years 1910–1914.[56] Local tribes revolted against the Japanese, including those in Wushe, a mountainous area in central Taiwan. During the infamous Wushe incident of 1930, historians are of the agreement that Japan used chemical warfare agents, in this case tear gas, in order to crush the rebellion led by tribal leader Mona Rudo.[57] (Japan began production of CW agents in 1928 on Okunoshima Island.) During the 1930 uprising, 134 Japanese were killed by aboriginal guerillas. In response, Japan sent an army and marine contingent of over a thousand men, with another 668 armed police. The Taiwanese aboriginals put up a desperate fight, using a combination of primitive firearms and hunting bows. In addition to employing co-opted aborigines as bounty hunters, Japan crushed the rebellion using "Green Canister" shells (chloracetophenone, or CN). In the end, 644 of the aboriginals were dead, representing about half of the indigenous community in Wushe. This particular engagement may have been part of ongoing field tests with CW, conducted by the Japanese on Taiwan between 1930 and 1941.

Ethiopia: 1935–1936

> For seven days without a break the enemy has been bombing the armies and people of my country, including women and children, with horrible gases. Hundreds of my countrymen are screaming and moaning with pain. Many are unrecognizable since the skin has burned from their faces.
>
> —*Princess Sehai (then 16), daughter of Ethiopian leader Haile Selassie, 1936*[58]

On October 3, 1935, the dictator of Italy, Benito Mussolini, invaded Abyssinia (modern Ethiopia). Heavily outgunned and ill-equipped to fight a modern war, Ethiopian forces put up a valiant, guerrilla-style struggle in their defense. In an attempt to achieve quick victory, Mussolini's army utilized chemical weapons during this brief conflict. Events that took place 20 years prior to this war had much to do with Italy's decision to use such weapons.

During World War I, Italian forces suffered two major chemical attacks. At the plateau of Doberdo in 1916 and in Caporetto in 1917, Italy was subjected to bombardment from German shells (launched using the improved projectors), containing chlorine-phosgene mixtures. (The subsequent humiliating retreat from Caporetto is described in Ernest Hemingway's *A Farewell to Arms*.[59]) In the 1930s this memory only served to intensify a keen interest in CW by the Italian military. During Field Marshal Pietro Badoglio's campaign in Ethiopia, riot control agents (RCAs) were first tried with uneven success. In the beginning of the war, Ethiopian resistance was scattered in the rugged countryside, and guerillas were often ensconced in caves. Only when the Ethiopian soldiers became more clustered and presented themselves as fixed targets did Italy attack using mustard in bombs and spraying devices.

In an evaluation of the war in Ethiopia, Major General Sir Henry F. Thuiller, former director of the Allied Gas Services in World War I, and J. F .C. Fuller, the English military theorist, agreed that mustard was an effective tool used by Italy to achieve quick victory. Furthermore,

> It also had the effect of reducing the total sufferings and loss of life. If the war had been prolonged to the following year, the advance to and capture of Addis Ababa would almost certainly have entailed heavy fighting, and this would inevitably have caused considerable casualties to the Italian forces. It may also be argued that such prolonged operations would have caused more losses and more wounds and suffering to the Abyssinians than were caused by the mustard gas.[60]

Diplomatic circles were quite severe in their recrimination of Italian CW in Ethiopia. The League of Nations initiated an embargo on Italy, excepting oil and steel. Moved by the Ethiopian example, British Foreign Secretary Anthony Eden spoke about it before the League of Nations in 1936:

> How can we have confidence that our own folk, despite all solemnly signed protocols, will not be burned, blinded, done to death in agony hereafter?[61]

WORLD WAR II

Most think of World War I as the conflict characterized by chemical warfare, and indeed this makes some sense, since it was in that war that many of the modern agents of poisoning, burning, and asphyxiation were used for the first time on a wide scale. But World War II and the conflicts building up to it had their share of chemical warfare as well, much of it in the Far East. And of course it was in the 1939–1945 conflict that chemical poisoning—on a scale never seen, and probably never even imagined—was visited on a huge civilian population.

The Sino-Japanese War

As far as is known, aside from an occasional cyanide grenade tossed at US soldiers (without apparent result), the battlefield use of chemicals in World War II was limited to the Sino-Japanese theater of operations. The Imperial Japanese Army employed chemical weapons against the Chinese during World War II (starting 1937 up until at least 1942).[62]

Reportedly, whole battalions of unprotected Chinese troops were routed by a combination of CW agents, from non-lethal harassing agents to phosgene and blister agents (mustard). During one particular engagement in 1940, Japanese troops used 300 kilograms (660 pounds) of mustard in an attack on Chinese communist forces at Shanxi province. General Tang En-po said this about the Japanese use of CW against his Chinese troops:

> ...Even when it is only tear or mustard gas, it lays our men out
> for long enough to enable the enemy to come and bayonet them
> as they lie gasping for breath....[63]

Though hardly in any position at the time to do much about it, in June 1943 President Roosevelt reiterated a threat to respond in kind to Japanese use of chemical weapons:

> ...Acts of this nature committed against any one of the United
> Nations will be regarded as having been committed against the
> United States itself and will be treated accordingly.... We prom-
> ise to any perpetrators of such crimes full and swift retaliation in
> kind.... Any use of gas by any Axis power, therefore, will imme-
> diately be followed by the fullest possible retaliation upon muni-
> tions centers, seaports, and other military objectives throughout
> the whole extent of the territory of such Axis country....[64]

Quoting a Soviet source, a book written by specialists in CBW defense for the Chinese People's Liberation Army claims that "during its war in China, the Japanese army had prepared 25 percent of their artillery shells to be chemical munitions, while 30 percent of its aerial ordnance were chemical bombs."[65] More precise if not accurate statistics from the same source record that from July 18, 1937, to May 8, 1945, Japan carried out 1059 chemical attacks in China, including use of the agents diphenylchloroarsine, diphenyl-cyanoarsine, chloracetophenone, chloropicrin, hydrogen cyanide, phosgene, mustard, and Lewisite. As Yu Zhongzhou, a Chinese arms control specialist, states, if Japanese chemical attacks caused nearly 100,000 casualties,[66] there should probably be more records of injuries from blister agents, for example, that would have been obvious to observers on the ground. There is a 1938 report from the Red Cross, signed by five physicians on the scene, reporting that at Xuzhou, "a large number of wounded soldiers was rushed to the hospital. Among them they found several cases showing generalized skin blisters and lesions resembling more or less those caused by smallpox... photographed evidence is available."[67]

While the use of chemical weapons probably assisted the Japanese in some battles and was able to lower their own casualty rates, it is difficult to conclude that CW played a decisive role in the outcome. Following Japan's surrender in 1945, the Kuomintang Garrison Command took control of former Japanese military facilities, including a "large chemical weapons facility in northern Taiwan."[68] It is unknown what stocks were found when Nationalist soldiers arrived at this plant.

As one might expect, the Chinese are bitterly indignant over Japan's use of CW in World War II, but also shifts some of the blame onto the United States. One Chinese source on the topic notes that, despite Roosevelt's warning to Japan in 1942 over their use of such weapons against the Chinese, the United States never did take measures to retaliate in kind.[69]

United States and CW Policy

During the hard-fought "island hopping" campaign in the Pacific, some military leaders and strategists in the United States considered the use of chemical weapons. General George C. Marshall recommended the use of CW in Okinawa, for example, but Secretary of War Henry Stimson rejected the idea. Throughout his public career, Franklin Roosevelt had been firmly against the idea of offensive chemical warfare,[70] in 1943 pledging not to initiate CW, and Secretary Stimson had no intention of violating that promise. Even the use of white phosphorus as an incendiary was under some debate by General Dwight D. Eisenhower and his advisors before the D-Day invasion. The United States was concerned that using white phosphorus—a substance that burns at a high temperature and generates large quantities of smoke—might violate the 1925 Geneva Protocol, which the US had signed but not ratified.

Churchill and Chemical Weapons

In the European theater, Allied and Axis powers considered the use of chemicals against one another, but neither did, fearing both retaliation in kind and moral condemnation from the world community. After intelligence reports indicated that Germany might be contemplating the use of chemical weapons against the Soviet Union in 1942, Winston Churchill broadcast the following message to the Reich:

With the goal of making children less afraid of its appearance, Walt Disney designed this gas mask for American children. (Courtesy of Soldier Biological and Chemical Command, Historical Research and Response Team, Aberdeen Proving Ground, MD.)

> ... I wish now to make it plain that we shall treat the unpro-
> voked use of poison gas against our Russian ally exactly if it were
> used against ourselves and, if we are satisfied that this new outrage
> had been committed by Hitler, we will use our growing air-supe-
> riority in the West to carry gas-warfare on the largest possible
> scale far and wide upon the towns and cities of Germany.[71]

On June 13, 1944, a week after the D-Day invasion at Normandy, Germany initiated an attack against London with its "revenge" weapon, the V-1 "Buzz Bomb." This was the first concerted use of unmanned aerial vehicles ever used for such a purpose. For 80 days, these bombs fell on civilian targets, killing over 6000 people. Churchill later remembered, "One landed near my home at Westerham, killing, by cruel mischance, twenty-two homeless children and five grownups collected in a refuge made for them in the woods."[72] In a state of vengeful fury, Churchill seriously considered the use of gas following the barrage. Many years before, he had successfully advocated for the use of chemical weapons in World War I against Germany and was quite prepared to do so again, and in writing. In his personal brief to the Chief of Staff, General Hastings Ismay, he laid out his arguments:

> ... If the bombardment of London really became a serious nui-
> sance and great rockets with far-reaching and devastating effect
> fell on many centres of Government and labour, I should be pre-
> pared to do anything that would hit the enemy in a murderous
> place. I may certainly have to ask you to support me in using poi-
> son gas. We could drench the cities of the Ruhr and many other
> cities in Germany in such a way that most of the population
> would be requiring constant attention. We could stop all work at
> the flying bomb starting points. I do not see why we should
> always have all the disadvantages of being the gentleman while
> they have all the advantages of being the cad. There are times
> when this may be so but not now. . . .[73]

But for all of the arguments made in favor of using chemical weapons against Germany, none were throughout World War II. The non-use of CW on the European battlefield is the subject of a good deal of scholarship. After all, by 1945, Germany had arguably the most advanced technology in the manufacture of nerve agents. Although Germany had seriously contemplated the use of chemicals in V-1 bomb and V-2 rockets, in the end the *Wehrmacht* generals concluded that conventional explosives were more efficient in terms of weight and destructive power than CW agents. A former officer of the *Wehrmacht*, Hermann Ochsner, told the Historical Office of the United States Chemical Corps why he believed Germany held back:

Designed to launch 150-mm rockets, the German Nebelwerfer *was first built in the early 1930s. Although Germany only used the* Nebelwerfer *to fire high-explosive artillery, it was capable of launching chemical ordnance as well.* (Courtesy of Soldier Biological and Chemical Command, Historical Research and Response Team, Aberdeen Proving Ground, MD.)

> ...The "V" weapons developed in Germany, V-1 and V-2, were
> not intended for gas warfare for the following reasons: the field of
> dispersion was too wide and the carrying capacity of the individ-
> ual projectile was too small, so that with the very low rate of fire
> it would not have been possible to gas any considerable area.
> Hence only locally-restricted and relatively small danger zones
> with gas coverage could have been created....[74]

There was another factor, however, in Germany's decision, and it involved a crucial misreading of intelligence. Although Hitler was tempted to use novel nerve agents such as tabun against the allies, his scientific advisers urged caution. They knew the scientific literature and also understood that chemicals called organophosphates—of which nerve agents are a member—had been investigated early in the nineteenth century by the Russian chemist Arbuzov at Kazan, Russia. Furthermore, the United States had classified much of its sensitive chemical reports since its entry into the war. Surely, Hitler's advisors counseled, the Allies must have discovered the toxic properties of tabun and similar compounds. It strained their credulity that Germany had a monopoly on tabun, sarin, or any other organophosphate agent. In point of fact, Germany actually did have a virtual monopoly on nerve agents, especially in their mass production. As it happened, the secret agent that the United States did keep classified was in fact DDT, a revolutionary insecticide that did make great contributions to winning the war by reducing large numbers of infectious disease among the Allies, but it certainly was no chemical weapon.

Perhaps Hitler's personal experience, having been injured himself by mustard in World War I, made him to hesitate before escalating to an aerial CW campaign. Or perhaps some Germans wished, in this case, to honor the code of German military officers—even in the *Waffen SS*—which precluded the use of chemical warfare. But that would ask us to attempt to understand a code of honor that allows for the genocidal murder by poison gas of millions of defenseless men, women, and children while at the same time prohibiting the use of the same gas against soldiers.

The Bari Incident

The production, transport, and disposal of chemical weapons created some environmental disasters both during and after World War II. The Bari incident of 1943 involved a spill of mustard agent off the southeastern coast of Italy. Large amounts of the agent—originally shipped to Allied forces in Italy for use in the event of a chemical attack by Germany—spilled into the water after an air assault by the *Luftwaffe* that sank several cargo vessels. The released mustard agent that floated on the oily surface of the water killed and grievously injured hundreds of Allied sailors who had been thrown into the sea.

FROM KOREA TO THE GULF WAR

The period after World War II has not been characterized by extensive use of chemical weapons, but they have certainly made their appearance on more than one occasion. As in the period after World War I, there has been a considerable effort to establish international covenants that prohibit the manufacture and deployment of these agents (see Chapter 9).

Allegations of Chemical Warfare in Korea

The People's Republic of China and North Korea alleged that the United States employed chemical and biological weapons during the Korean War. Recently discovered documents show that at least the biological warfare charge was based on a North Korean disinformation campaign.

While some commanders in the US armed forces considered the use of chemical weapons in the Korean War, no credible evidence has been found to support subsequent allegations of CW being used against the North Korean and Chinese armies. Furthermore, there is circumstantial evidence that is exculpatory.

Five years after the end of World War II—when US defense spending and preparedness was at its low point—military forces sent from the United States and elsewhere were not equipped to handle CW offensively. Nor is it likely that chemical weapons would have been of much help at the Pusan perimeter, for this could have resulted in Soviet-supplied chemical weapons being introduced by the North.

During the latter stages of the war, a massive US retreat from attacking Chinese forces (otherwise known as the great "bugout") left both dead American soldiers and tons of materiel behind. Had chemical weapons been brought to the theater, no doubt the Chinese or North Koreans would have found it. In any event, neither Beijing nor Pyongyang ever provided any physical evidence supporting the CW charges. Because the member states of the United Nations were so vocal in their disapproval of CW, it is highly unlikely that such activity would have taken place without foreign nationals discovering the program and dissolving the coalition.

Perhaps the most dramatic testimony that controverts the Chinese CW allegations of the Korean War was later found in the Soviet archives. Lt. Gen. V. N. Razuvaev, former Soviet Ambassador to North Korea and military advisor for the Korean People's Army, wrote the following to Levrenti Beria on April 18, 1953:

> . . . the Chinese . . . wrote that the Americans were using poison
> gas in the course of the [Korean] war. However, my examinations

into this question did not give positive results. For example, on April 10, 1953, the general commanding the Eastern Front reported to Kim Il Sung that 10–12 persons were poisoned in a tunnel by an American chemical missile. Our investigation established that these deaths were caused by poisoning from carbonic acid gas [i.e., CO_2] [released into] the tunnel, which had no ventilation, after the explosion of an ordinary large caliber shell.[75]

In retrospect, the Russian archives explain best why Chinese military leaders could have believed that the United Nations armies were using chemical warfare. The Chinese People's Volunteer Forces faced terrific air power later in the Korean War, as well as US-delivered ordnance such as napalm, artillery, and aerial strikes using fighter/bomber aircraft. In addition to the immediate, devastating effects of these attacks, resultant off-gases from bombardments were no doubt responsible for respiratory distress and pulmonary edema among Chinese soldiers, symptoms that are largely indistinguishable from lung irritants found in chemical weaponry.

Yemen: 1963–1967

When Egyptian leader Gamal Abdel Nasser intervened militarily during the Royalist and Republican civil war in Yemen, chemical weapons were employed, perhaps as early as 1963. Journalists who were present in Yemen in 1967 reported a large chemical agent attack that killed over 100 people. There is also an unconfirmed story that Egypt also used a nerve agent (sarin, or GB) in Yemen, which would make it the first such recorded use. However, Milton Leitenberg reports that at least some of these reports were actually organophosphate pesticides (such as parathion), filled in jerry-rigged canisters attached to hand grenades.[76] While accounts have averred that Egypt utilized leftover British stocks of chemicals, the discovery of bomb fragments with Cyrillic-typed labels pointed to their Soviet origin.

Southeast Asia: 1965–1975

A now thoroughly discredited story appeared in the western press (*Time*/CNN) in June 1998, alleging the use of the nerve agent sarin (GB) against Vietnamese, Laotians, and even suspected US defectors during the Vietnam War. No evidence has been since provided that can support the charge, and nearly all of those US personnel cited as sources for the story—including key participants in "Operation Tailwind"—have either denied, discounted, or refuted the story altogether.[77] It is most probable that the agent in question was in fact a riot control agent, namely CS. Other accounts of

Vietnam War-era reporting also lead to the conclusion that CS tear gas, a non-lethal compound, was used in at least that particular operation.[78]

Iran-Iraq War: 1980–1988

During the Iran-Iraq War, Iraq employed chemical warfare both as a battlefield and a terror weapon against civilian (especially Kurdish) targets. After a series of rapid Iraqi successes in capturing Iranian territory early on, Iranian forces—many made up of young *bassij* volunteers—used human wave attacks, as well as some creative battlefield and riverine tactics, to drive Iraq back to *ante bellum* lines. In the end, however, Iraq prevailed against Iranian forces, and Iran was forced to sue for peace.

On August 12, 1981, Iran first reported that Iraq was using CW agents, and while it is conceivable that Iraq was in fact experimenting with chemical weapons this early in the war, the allegation has not been confirmed. But by March 1984, the United Nations did confirm the use of chemical weapons—including mustard and GA (tabun) nerve agent—by Iraq. Sheer desperation, however briefly it may have lasted, also forced Iraq to use CW agents in the spring of 1987 when it appeared that Iran could win by a war of attrition.

Iraq's learning curve with respect to chemical weaponry included some fitful starts. Early efforts at employing CW agents led to Iraqi self-inflicted casualties. Throughout the conflict, there was a failure to use massed amounts of CW agents for decisive effect. But by 1987, probably with the help of technical advice from the Soviet Union, Iraqi military commanders made significant improvements in CW operational art. Iraq began a sophisticated approach to attacking with CW agents, delivering simultaneously non-persistent nerve gas on forward positions and persistent mustard agent against the enemy's rear logistical areas.

Earlier in the conflict, Iran repulsed the initial Iraqi invasion and even made further gains, taking some Iraqi territory during 1982–1984. But not only was Iraq better armed, it had substantial advantage in logistics. Iran, therefore, used large numbers of personnel in extremely costly offensives to make up for their own deficiencies in these areas. Using a revolutionary battle strategy reminiscent of Mao Zedong's People's War, poorly trained and often on foot, Iranian troops were extremely vulnerable to casualty agents such as mustard.

Though perhaps not critical to Saddam Hussein's victories at a strategic level, tactically the employment of nerve and blister agents was extremely effective against Iranian soldiers, especially during the human wave attacks by fanatical *bassij*. Although Iran did respond sporadically with chemical weapons later in the conflict, Iran clearly got the worst of it. Even so, CW accounted for only 3–5 percent of Iran's casualties during the eight-year-long war.

An Iraqi 500-kg mustard bomb. The note on the munition reads "polymerred," a reference that is unclear. It could mean that the munition used unique materials, or that the mustard fill was thickened with a polymer. This and other munitions were often found to be leaking at their storage sites. (Courtesy of United Nations, Photograph by H. Arvidsson.)

Agents Used Against the Kurds by Iraq

Chemical agents were used by Saddam Hussein's troops against Kurdish-populated villages in Iraq, including Halabja, where in 1988 at least 4000 civilians perished from gas poisoning. Iranian physicians reported that victims of the chemical attacks on Halabja showed characteristic symptoms of cyanide poisoning, while other reports indicated large amounts of mustard and other chemical weapons.[79] Reportedly, survivors of this particular attack have permanent injuries, including burns, and some exhibit symptoms of neurological damage, although this cannot yet be adequately confirmed.[80]

At the time of the attack, Iran was able to gain at first what seemed an extraordinary propaganda win, as it showed the world film footage of the aftermath of the Halabja massacre perpetrated by Iraq. Ironically, in the course of making its case, Iran had simultaneously made its own people publicly aware of Iraqi chemical weapons, including the horrors they can cause. The demoralizing effect of showing Iranians the aftermath at Halabja may have helped contribute to Iran's decision to sue for peace that same year.

In October 1992, samples brought from the Kurdish village of Birjinni in the Iraqi northern mountains were tested with highly sensitive laboratory techniques. Clothing, detritus, and munition fragments were collected, showing evidence of an Iraqi chemical attack against Birjinni four years earlier. Scientists at the Chemical and Biological Defense Establishment, in Porton Down in the United Kingdom, detected the presence of sarin (GB) nerve agent as well as sulfur mustard and other products.

Iranian Chemical Weapons Development

Chemical and biological weapons are the poor man's atomic bombs and can easily be produced. We should at least consider them for our defense. Although the use of such weapons is inhuman, the war taught us that international laws are only scraps of paper.

—former Iranian leader Hashemi Rafsanjani,
October 1988[81]

Iran used captured Iraqi chemical artillery shells against Iraq in 1984, but it was only much later in the war that Iran was able to respond with indigenously made chemical weapons. In 1986, an opposition newspaper reported that Iran finally got serious by increasing research funding for the *Jahad-e Daneshgahi*, or University Crusade, consisting of about 200 graduate students "carrying out various experiments on such products as mustard gas and military equipment."[82] Iran was capable of making extensive use of chemicals, notably during fighting around Basra in the spring of 1986, and by early 1988

Long-range Scud vehicle (modified "Al-Hussein") being destroyed by an
UNSCOM team in 1995. Scud warheads were filled with chemical and bio-
logical agents by Iraq. (Courtesy of United Nations, Photograph by
H. Arvidsson.)

was able to produce a full range of CW agents, including nerve, phosgene, and hydrogen cyanide.

Lessons from the Gulf War

When Saddam Hussein invaded Kuwait in 1990, he did not expect that the United States would respond as resolutely as it did—a serious miscalculation on his part. Nonetheless, the United States—faced as it was with an enemy that had used chemical weapons against Iranian forces and Iraqi Kurds alike— was extremely worried that Iraq might use chemicals against coalition forces during Operation Desert Storm.

For one thing, the US military realized it had an inadequate supply of chemical agent monitors and alarms. A hurried requisition brought thousands of Chemical Agent Monitors (CAMs) from England that could detect the presence of mustard. False alarms (some being caused by vehicle exhaust, for example) were all too common in some of the older US-made monitors, and soldiers, tired of being jolted by a warning that turned out to be false, turned off their chemical alarm equipment. With design requirements set by the US Department of Defense, newer chemical agent detectors were and are currently being built to be more sensitive and accurate.

Had Iraq used chemicals in its SCUD missile launchers or used artillery-based chemical munitions against US and coalition forces, it is very likely that a devastating response from conventional explosives (or even nuclear warheads) would have been the result. It is then quite likely that few of Iraq's Republican Guard assets would have survived the war. Perhaps because of this threat of coalition response, Saddam Hussein finally elected not to use CW.

Still, in preparation for the worst, US air and ground forces were equipped with protective garments. Apache helicopter pilots, in particular, had to deal with an inadequate air conditioning system—as well as heat-trapping chemical suits—all the while conducting operations in a very hot environment. Soldiers were supplied with antidotes that had a combination of atropine and another compound to reverse the effects of nerve agents. They were also prescribed pyridostigmine bromide (PB) tablets to take as a precaution to protect themselves from sarin or other nerve gases. The doses allowed were relatively small, but they did cause some discomfort among soldiers that did take them (intestinal gas and urinary urgency), and there were a few soldiers who elected not to follow this regimen.

Some have suggested that PB tablets, in combination with other factors, could have been a cause of Gulf War Syndrome (GWS), but little evidence supports this theory. As of early 1999, claims that the syndrome among coalition veterans is the result of chemicals in the conflict remain unproven.

Khamisiyah and Sarin Release in the Gulf War

No evidence to date has proven that Iraq deliberately used chemical or biological weapons against coalition armies during the Gulf War. However, some soldiers in the US, United Kingdom, and French ground forces might have been exposed to very low levels of nerve agent in the last stages of the conflict.

The United States Central Intelligence Agency was aware of the Iraqi Khamisiyah arms depot as a possible chemical weapons storage site, but word did not reach US commanders in the field when forces were sent to destroy the depot in March 1991. (Apparently there was some confusion by analysts concerning the exact name, location, and nature of the depot.) Only in June 1996 did the US Department of Defense become convinced that Khamisiyah had contained more than 8 metric tons of sarin before being blown up by the US Army.

When the site at Khamisiyah was demolished with explosives, the blast released sarin nerve agent in a plume. However, the levels of agent in the air would have been of very low concentration, and none of the personnel known to be in the immediate vicinity showed any symptoms of nerve agent exposure. Furthermore, computer models generated by the CIA were probably overestimating the original amount of nerve agent involved, as well as the size of area covered by the plume.

CHAPTER 6

Control and Disarmament

Historically, chemical warfare has sometimes been viewed at the very least with suspicion, and more often with a combination of terror and abhorrence. Chemical weapons were considered, and remain in the minds of most of us, uncivilized if not barbaric tools of war. As early as 1675, France and Germany both condemned the use of poisoned bullets in combat, and over the years, dozens of international treaties and declarations have been drafted to ban or limit their use. In this chapter we take a look at some of these attempts at disarmament and control, and speculate on not just the chances but even the desirability of treaties.

HISTORICAL PRELUDES

There were numerous efforts to control the spread of chemical weapons prior to World War I, but most, as that war amply demonstrated, were ineffectual. In particular, two conventions, the First Hague Conference (1899) and the Second Hague Conference (1907) went a long way in establishing certain rules of war (in respecting the neutrality of ships on the high seas and in protecting noncombatants, for example). And the conferences left a legacy that is in place to this day—the court we know as the Hague Tribunal. But in terms of reducing the proliferation and stockpiling of armaments, including poison gas, nations were for the most part left to their own devices.

Early Twentieth Century Negotiations

As Austria grew in military might at the end of the nineteenth century, Russia, then in the midst of building a far-reaching railroad network, was unable to keep up with its own defense spending. In 1898, Tsar Nicholas—motivated as much by *realpolitik* as by an authentic desire for peace—proposed a convention at which participant nations would agree to limit armament growth.

The Hague Conferences

Although there was international skepticism, twenty-six nations accepted the Tsar's invitation, and the group convened for the first time at The Hague, in the Netherlands, in 1899. Officially named the International Peace Conference of 1899, and the second meeting in 1907, known respectively as the First Hague Conference and the Second Hague Conference, became models for international negotiation and cooperation—at least in the high-mindedness of their intentions.

Although no wide-reaching disarmament agreements emerged from the two conferences, three important declarations regarding chemical warfare were made. Reflecting sentiments expressed in the Brussels Declaration of 1864, as well as the Brussels Convention of 1874, the first conference called for a prohibition on the use of poisons or gases, as well as other new technologies, that the assembled nations believed would cause unnecessary suffering in war. One declaration made in 1899 banned the use of projectiles from balloons, another eliminated the use of "dum-dum," or expanding bullets. And the third prohibition concerned itself with certain projectiles, "the object of which is diffusion of asphyxiating or deleterious gasses."[1] Although the larger aim of the meetings, to limit armament growth in Europe, was not achieved, the delegates did begin to define the acceptable parameters of modern warfare.

Even though many of the delegates to the First Hague Conferences seemed committed to eliminating or at lease restricting these new weapons of warfare, voices of dissent were raised. Admiral Alfred T. Mahan,[2] the American delegate, was not convinced that gassing troops was inherently more evil than drowning sailors in naval engagements:

> . . . It is illogical and not demonstrably humane to be tender about asphyxiating men with gas, when all are prepared to admit that it is allowable to blow the bottom out of an ironclad at midnight, throwing four or five hundred men into the sea to be choked by the water, with scarcely the remotest chance to escape. . . .[3]

Ultimately, the United States chose not to sign the "Final Act," the July 29, 1899, umbrella document covering all the declarations that emerged from the 1899 Hague Conference. Others, Germany included, signed and ratified most of the declarations, including the pledge to "Abstain from the use of projectiles the sole object of which is the diffusion of asphyxiating or deleterious gases." However, an important qualifier was added to this declaration:

> . . . that the regulation shall be binding upon the powers only in case of war between two or more of them, and *shall cease to be binding in case a non-contracting power takes part in the war*. . . .[4]

The years following the First Hague Conference saw great advances in aviation, in the design of land mines, and, as discussed in Chapter 5, the science of organic chemistry. Predictably, by the time the next round of meetings concluded in October of 1907, many nations were much more willing to seriously consider limits on a host of the new tools of war, including chemical agents. Added to the language of one of the conference's conventions was a terse and unambiguous admonition about the rules for war on land: "It is expressly forbidden . . . to employ poisons or poisonous weapons."[5]

Despite the high-mindedness—the faith in negotiation and enlightened self-interest and international cooperation—the declarations set forth at the Hague Conferences stopped not a single one of the major powers from developing chemical weapons. Within a few short years, they were all putting them to use on the battlefield. Even though the Germans, for example, had been an active and willing participant at the meetings, they were the first to use chemical weapons, less than ten years after the second conference ended. By the end of the war, approximately 124,000 tons of chlorine, mustard, and other chemical weapons had been released, virtually all of it by signatories to conventions issued at the Second Hague Conference.

The German government and public, during and after World War I, did not shy away from making the argument that chemical weaponry was actually a *humane* form of combat. In June 1915, the editors of a Cologne, Germany, newspaper opined that

> . . . The basic idea of the Hague agreements was to prevent
> unnecessary cruelty and unnecessary killing when milder methods of putting the enemy out of action suffice and are possible.
> From this standpoint the letting loose of smoke clouds, which, in
> a gentle wind, move quite slowly towards the enemy, is not only
> permissible by international law, but is an extraordinarily mild
> method of war. . . .[6]

In 1927, Russel H. Ewing, an apologist for the use of chemicals in war, argued in *The American Law Review* that chemical weapons were permitted in World War I because two belligerents, Turkey and Serbia, never ratified the 1907 Hague Convention prohibiting gas warfare. The same article went on to reason that Germany did not violate the clause regarding the use of "projectiles, the sole object of which is the diffusion of asphyxiating or deleterious gases. . . ," because in the German's first major chlorine attack, there were not projectiles but static tanks and cylinders. Finally, the scholar noted, when chemicals were used in tandem with explosive projectiles and bombs, technically the delivery of the chemical component was not the "sole object."[7]

These tendentious legal arguments aside, the 1919 Treaty of Versailles, in addition to its many punitive measures levied against a defeated Germany, further was used by the Allies to prohibit Germany from acquiring chemical weapons. This added measure did as much good as the Versailles Treaty with regard to restricting the re-arming of Germany in other areas, which is to say nothing.

The Washington Arms Conference

In the years following World War I, the international community attempted once again to limit the use of chemical weaponry, this time at the Conference for the Limitation of Naval Armament in 1921 in Washington, DC. Amidst deliberations were other, more pressing issues such as "capital ships," tonnage and displacement of naval vessels. The American delegation to the meeting (later referred to as the Washington Arms Conference of 1922) presented polling data showing that the American public overwhelmingly supported abolishing chemical weapons. Although chemical weapons were not purposely used against civilians in World War I, the horrific casualties that they had wrought shocked and terrified the American people. The survey found 367,000 in favor of making CW obsolete—and 19 in favor of its retention.[8]

While public opinion may have helped to persuade the American delegation, chaired by General John J. "Blackjack" Pershing, to argue for the prohibition of chemical weapons, the American delegates had a number of other concerns. First, they knew that CW gases could deal a tremendous blow against unprepared armies, and since the United States was far behind Europe in research on and development of these agents, the Americans did not want to risk entering into an agreement in which a technically advanced but unscrupulous nation would have a significant battlefield advantage over US forces. Second, since many high explosives also produced gases that had a chemical weapon-like effect on soldiers, any attempt to forbid the use of CW gases would lead to confusion, since it would be difficult if not impossible to distinguish between casualties that resulted from high explosives and those that were caused by chemical weapons. The Americans feared that an adversary could exploit this "gray area," using it as an excuse to launch a heavy attack with CW gases in every form.[9] And third, the Americans were concerned there were no restrictions on research, which could lead to the discovery of completely new warfare gases.

The American delegation did, however, note the need for further discussion concerning the beneficial aspects of gas warfare, finding as follows:

> . . . that there are arguments in favor of the use of gas which ought to be considered. The proportion of deaths from their use, *when not of toxic character*, is much less than from the use of other weapons of warfare. . . .[10]

Finally, after much debate, the US Senate unanimously ratified the agreement, with only minor objections made during floor debates. But in the end, French objections on treaty language concerning the use of submarines scuttled the agreement, and no agreements from the Washington Arms Conference of 1922 ever went into force. There were many reasons for the ultimate failure of the Conference, but even had some treaty prohibiting chemical warfare been ratified, only the United States, France, Great Britain, Italy, and Japan as the actual parties to these disarmament talks would have signed. Like the later 1927 Kellog-Briand Pact that aimed to outlaw *all* warfare, the 1922 Conference suffered from the naïve mantra that merely outlawing the weapons of war would somehow stop further conflicts. With convoluted rules and ratios for naval vessels among the major powers at the time, and even an attempt to prohibit the use of submarines, the efforts in 1922 were at best a waste of time. At worst, they actually helped precipitate the start of World War II.

Yet another attempt to prohibit the use of chemical and biological weapons would take the form of the Geneva Protocol, but it too had very little success in stemming the development of this kind of warfare.

The 1925 Geneva Protocol

By 1928,[11] most of the world community, by its adoption of the Geneva Protocol of 1925, had—at least on paper—foresworn the use of gas in war. Originally intended to stem international trade in conventional weapons, the Geneva Protocol called for the prohibition of "asphyxiating, poisonous or other Gases, and bacteriological methods of warfare."[12] Though farsighted in the sense that it banned biological as well as chemical agents, the Protocol also contained key conditions and exemptions that made it a very ineffective treaty. For while the agreement prohibited the use of such weapons in battle, it did not expressly prohibit countries from developing, producing, and stockpiling chemical and biological weapons arsenals. In addition, it did not restrict the use of chemical and biological agents against non-ratifying parties, making it a *de facto* "no first use" agreement against any country that had not signed the agreement. The Protocol also permitted the use of chemical weapons in civil conflicts within a country's own borders.

By the time of the 1925 Geneva Conference, the US State Department wanted a ban the export of war gases and supported that position at the conference. Joining the Americans, France wanted to ban the chemicals entirely, and Poland recommended that the agreement further prohibit bacteriological warfare as well. And yet, even as American delegates pushed their case in Geneva, a campaign led by the American General Amos Fries, chief of the Army Chemical Warfare Service, argued that if the Senate ratified the

Major General Amos A. Fries was chief of the US Army Chemical Warfare Service (CWS) during the 1920s. With determination and considerable political acumen, Fries ensured the survival of the CWS. (Courtesy of Soldier Biological and Chemical Command, Historical Research and Response Team, Aberdeen Proving Ground, MD.)

Protocol, America would find itself at a significant disadvantage in military readiness. (The American Chemical Society also weighed in against the agreement.) In the end, the United States did ratify the Geneva Protocol—but almost half a century later, in 1975.

THE CHEMICAL WEAPONS CONVENTION (CWC)

During the Cold War, both the Soviet Union and the United States were reluctant to consider any type of disarmament. Even as many nations continued to develop and stockpile chemical weapons, and as new players undertook to arm themselves with the agents, nuclear weapons took center stage. The world of disarmament was dominated by test-ban treaties and nuclear nonproliferation pacts.

As a consequence, the first significant discussions of a ban on chemical weapons since the Geneva Protocol of 1925 started soon after the conclusion of the Biological and Toxin Weapons Convention (BTWC) in 1972. During negotiations at UN disarmament conferences in Geneva, exploratory initiatives by Japan (1974) and the United Kingdom (1976) were proposed for a chemical weapons ban.

In the discussions that followed, and that continued well into the 1980s, negotiations hinged on the matter of "intrusive verification"—which in arms-control lingo roughly means on-site inspections. Not surprisingly, intrusive measures of this kind were not accepted by the former Soviet Union and her allies. By 1987, however, the main protagonists in the discussion—the Soviets and the US—had built up enough of a framework of trust to agree to mutual "challenge inspections." This type of weapons-verification procedure is especially intrusive, allowing one party to select and enter another party's site on short notice and with minimal guidance and control. Unimaginable ten years earlier, by the end of the 1980s the world's two largest producers of chemical warfare agents were looking into each other's stockpiles. Although debates over the issue of inspections continued for years, by 1992 an agreement was finally reached. The CWC was finally opened for signature on January 23, 1993, and on April 29, 1997, six months after the sixty-fifth country submitted its "instrument of ratification" to the Secretary General of the United Nations, the agreement was in force.

Although debates over the issue of CW inspections and disarmament continue to this day, 1992 marked the starting date of a truly historic ongoing agreement. Having learned the hard way from the deficiencies of the Geneva Protocol of 1925, the CWC was written more forcefully, and refined almost endlessly, in the hopes that chemical weapons would be banned, and banned in a comprehensive and permanent fashion.

First, the CWC prohibited the "development, production, acquisition, retention, stockpiling, transfer and use of all chemical weapons."[13] Furthermore, states that signed and ratified the agreement were required to declare any and all CW stockpiles. They were also required to destroy their chemical weapons within a reasonable amount of time—in anywhere from five to ten years—depending on their specific circumstances. States were also held responsible for the clean-up of any chemical weapons that had been abandoned in another state's territory. In addition, CWC members agreed to destroy or convert to peaceful use any chemical-weapons production facilities operated since 1946 under their jurisdictions.

Of course, the sweeping directives of the convention eventually had to be applied to the countless details of the real world, and in the details was the devil. Some chemical factories in the former Soviet Union, supposedly adaptable to civilian use, were, in the eyes of western inspectors, still potential weapons factories. The Russian authorities, faced with a collapsing economy, resisted and are to this day resisting the outright destruction of certain facilities. And interesting exceptions were made to the ban—the manufacture and stockpiling of riot-control agents were permitted for domestic law-enforcement purposes, and in a few other cases the United States reserved the right to hold on to chemical agents that were otherwise banned "as a method of warfare."[14]

Nonetheless, the CWC is a remarkably sturdy and effective agreement, both demanding of its signatories as a group and flexible enough to meet individual states' needs. And the convention was written to encourage participation: membership is open, and any state is allowed to join, regardless of whether it possesses chemical weapons. Unlike the Nuclear Nonproliferation Treaty (NPT) of 1970, which treats nuclear powers differently from those considered non-nuclear states, the CWC has no special requirements, provided the new member state accepts the same degree of transparency and intrusive verification required of other members.

Complying with the CWC is challenging, even for the wealthiest member states. Possessing between them approximately 70,000 tons of chemical agents, the United States and Russia will be destroying their stockpiles well into the next decade, and perhaps until 2020. And Japan, for example, has agreed to fund the disposal of at least 680,000 chemical shells that were abandoned in China during World War II. Still, the member states have, for the most part, continued to abide by the convention's general principles and individual directives.

Controlling Agents and Precursors

It is critical to note that not all compounds are treated similarly under the CWC, which makes very clear distinctions based on the relative threat of each substance. First, any compound that can be used in combination with other compounds to form a CW agent is called a "precursor" and is regulated by the CWC. The CWC then lists both CW agents themselves and their precursors in three "schedules," or categories. Precursors are placed into schedules depending upon their immediate potential for being used as a weapon, and the relative ease of conversion into a CW agent. The CWC also allows for the fact that many precursors are used for commercial industries and has tried to make reasonable allowances for this by regulating these compounds less stringently.

Under the CWC, toxic chemicals and precursors can only be acquired for peaceful purposes, although a "State Party"—a nation participating in the Convention—can justify producing or obtaining toxic substances in order to develop defenses against them. Devices such as bombs that are specifically designed to deliver CW agents are also banned under the CWC.

Scheduling Agents and Precursors

The following examples illustrate the purported logic behind the placement of a chemical in Schedules 1, 2, or 3.

Schedule 1 Agents and Precursors

Schedule 1 consists of chemicals that are known CW agents—substances that could be used as weapons and have little or no use in commercial industry. Schedule 1 chemicals include:

- Sarin, soman, tabun, VX (nerve agents)
- Sulfur and nitrogen mustards,[15] and their analogues
- Lewisite and its derivatives
- Ricin
- Saxitoxin

Each of these substances is tightly monitored; in fact the CWC dictates that:

> The national aggregate of all Schedule 1 chemicals within a State Party may not exceed 1 ton at any given time. Production data must be declared for the following: the single small-scale facility (SSSF) with a maximum annual production of up to 1 ton per year per State Party; for protective purpose facilities with aggregate production of up to 10 kg per year per State Party; and for research, medical, and pharmaceutical facilities with an aggregate

production of up to 10 kg per year per facility. Any transfer of Schedule 1 chemicals between States Parties must be declared in advance by both States Parties. All declared facilities will be subject to routine verification inspections.

—*Organization for the Prohibition of Chemical Weapons (OPCW), The Hague: "A Guided Tour of the Convention on the Prohibition of the Development, Production, Stockpiling and Use of Chemical Weapons and on Their Destruction."*[16]

While most of the Schedule 1 compounds do not have commerical uses, nitrogen mustard (HN-2) has been used in chemotherapy, and both ricin and saxitoxin have demonstrated legitimate roles in medicine. Otherwise, there are no legitimate or even plausible uses for significant quantities of any Schedule 1 compounds. Schedule 1(B) precursors include compounds that readily form CW agents, such as the binary components for nerve agents such as QL for VX, and chemicals that are a simple step away from sarin and soman (for example, DF or methylphosphonyldifluoride). Precursors in Schedule 1 are so close structurally to being nerve agents that only a minor processing step, or the use of a common reagent, would result in a prohibited end product.

Schedule 2 Agents and Precursors

Schedule 2 chemicals are toxic in and of themselves. Some can serve as precursors of Schedule 1 CW agents, and most are only produced for commercial use in limited quantities. Toxic compounds listed under the CWC's Schedule 2 include:

- Amiton
- PFIB[17]
- BZ

Amiton, an organophosphorus pesticide that is especially toxic to humans, could be used as a weapon, especially in its pure (or "technical") form. Another, PFIB, ("p-fib") is the cause of "polymer fume fever" produced by the overheating of Teflon. The United Kingdom first brought the issue of PFIB during the 1989 Conference on Disarmament, which led to its official listing on CWC Schedule 2. PFIB is not a classic CW agent, but is considered a potential weapon due to its extremely high toxicity (possibly ten times that of phosgene). BZ, a very potent psychoactive incapacitant, is the third CW agent listed under Schedule 2.

State parties are only allowed to possess small quantities of these compounds, 100 kilograms (220 pounds) for Amiton and PFIB, respectively, and 1 kilogram (2.2 pounds) of BZ. Facilities that produce over and above these thresholds must make detailed declarations and are subject to CWC inspection. Schedule 2 precursors include those that can contribute to the manufacture of BZ, mustard, arsenic-based compounds (for example, Lewisite), and nerve agents (Schedule 1). For precursors, member states can produce one ton per year before declarations are required.

Schedule 3 Agents and Precursors

Schedule 3 chemicals are either potential CW agents themselves, or can be utilized as precursors for Schedule 1 and 2 compounds. Significantly, Schedule 3 precursors also have widespread uses in legitimate industry. Therefore, as long as they are "produced for purposes not prohibited under this Convention," Schedule 3 precursors "may be produced in large commercial quantities."[18] Schedule 3 toxic agents include:

- Phosgene
- Cyanogen chloride
- Hydrogen cyanide
- Chloropicrin

In addition to the classic CW agents listed above, Schedule 3 precursors include 13 compounds that can be used to manufacture nerve agents and mustard.

The general guidelines for Schedule 3 are as follows:

> Production, import and export data as well as the manufacturing sites must be declared for plants producing Schedule 3 chemicals in excess of 30 tons per year. State Parties must declare the national, aggregate amounts of each Schedule 3 chemical produced, imported, and exported, as well as quantitative data on the imports and exports for each country involved. If the production of a given plant exceeds 200 tons annually, CWC may conduct routine verification inspections of declared plant sites. Routine inspections will begin in the first year after entry into force of the Convention.
>
> —*Organization for the Prohibition of Chemical Weapons (OPCW), The Hague: "A Guided Tour of the Convention on the Prohibition of the Development, Production, Stockpiling and Use of Chemical Weapons and on Their Destruction."*[19]

Declarations and the CWC

Within a given time frame, CWC member countries must declare all of their chemical weapons production facilities, all CW stockpiles and storage facilities, any chemical weapon munitions (filled or unfilled), and all abandoned CW agents and munitions.

Significantly, a state that is party to the Convention has primary responsibility for the disposal and clean-up of chemical weapons abandoned on another state party's territory. As the verification annex of the CWC states, "[the] Abandoning State Party shall provide all necessary financial, technical, expert, facility as well as other resources. The Territorial State Party shall provide appropriate cooperation."[20] This is the clause that, for example, obliges Japan to survey the remains of chemical weapons its armies left behind in China and pay the associated costs of disposal.

Export Controls

Under the US Export Administration Act of 1979, the President can control the export of chemicals or equipment from the United States that "would assist the government of any foreign country in acquiring the capability to develop, produce, stockpile, deliver, or use chemical or biological weapons."[21] Furthermore, it is the policy of the United States to work in cooperative efforts to similarly curb the export of chemical weapon precursors internationally. Such efforts include an informal forum of nations that are significant exporters of chemicals and equipment, the Australia Group.

The Australia Group

The Australia Group is an international organization made up of 33 different countries trying to stem the proliferation of the chemical and biological weapons. The group was formed in 1984 in response to information that chemical weapons were being used in the Iran-Iraq conflict, and that international suppliers had sold the chemicals and equipment that aided both countries' CW programs. Similarly, after discovering the details of the Libyan CW program in 1989, and learning that western countries supplied Libya with necessary equipment, the Australia Group decided to begin monitoring chemical manufacturing facilities and technology.

The Australia Group aspired to both harmonize export controls on precursors and production equipment and share information that might stem the proliferation of chemical weapons. Initially, the Australia Group began with a list of 8 chemicals it thought were most prone to be used to make chemical weapons. Later, 46 more were added to make a current list of 54 controlled chemical exports. Many of these compounds are not listed in the CWC, but are nonetheless considered "dual-use" substances that have the potential to proliferate chemical weapons. The Australian Group countries apply measures to license dual-use materials and equipment that could be used in the manufacture of chemical weapons. To further increase awareness and cooperation between the chemical industry and lawmakers, the Government-Industry Conference against Chemical Weapons was formed, holding its first meeting in Canberra in September 1989. Today, the group meets biannually in Paris to discuss and agree upon measures to control the export of materials and technologies relevant to chemical and biological warfare. At this time, member nations also exchange data on CBW proliferation and consider means of implementing and expanding export controls.

For all of their good intentions, the Australia Group and, for that matter, the CWC itself face many challenges as they attempt to stem the proliferation of chemical weapons. First, most precursors have legitimate commercial uses, and this dual-use nature impedes detection of CW programs. If one were to count every chemical facility worldwide capable of producing CW agents or their precursors, the number would reach well over 10,000. And there are literally hundreds of facilities producing Schedule 2 compounds. Clearly, it will be tremendously challenging to verify compliance, even for those countries that are cooperative.

A nation that aims to undercut international agreements, and specifically that aims to undercut the CWC, will have a more difficult time obtaining chemical weapon precursors. However, some companies and nations can and do falsify documents and give misleading information about the final destination of chemical weapon components. When in 1984 Libya was building its CW agent manufacturing facility, Phara 150, Imhausen-Chemie, a German chemical firm, shipped the plans and components for a complete chemical plant to a firm in Hong Kong. In addition to serving as a holding company for the goods and plans, the Hong Kong firm held a controlling interest in a number of shipping companies that were subsequently used to transport equipment to Libya.

As long as companies are willing to sell controlled items and conceal their activities using false shipping manifests and front companies, it will be possible for a nation to acquire the materials and technology needed to produce chemical weapons. The only question is how much money and effort a government is willing to spend to acquire this capability—and how long it is willing to wait.

After having been caught trying to purchase precursor chemicals abroad, some countries, such as Iraq, created the main ingredient for mustard agent production domestically and used locally-mined phosphorus to manufacture nerve agents. Though an expensive undertaking, countries like Iraq could pay the premium for continuing their chemical weapons development by diverting large revenues from oil production.

Verification of Compliance

Those nations that have signed and ratified the CWC are considered State Parties to the convention, and collectively make up the Organization for the

Prohibition of Chemical Weapons (OPCW), based in The Hague, Netherlands. The Technical Secretariat of the OPCW is tasked with overseeing verification measures and handling the administrative details. As of January 2000, all of the 60 declared chemical weapons facilities "have been inspected and sealed," 20 are confirmed destroyed, and 3 approved by the OPCW to be converted for civilian use.

Without verification, the CWC participants would not be assured that other states are adhering to the agreement. In fact, the major strength of the CWC is its extensive provisions for intrusive verification. The process of verification is fraught with challenges. In addition to the weighty matter of sovereignty, there is the real or perceived threat that industrial secrets may be revealed during an inspection of a facility. Article VIII of the CWC attempts to allay these fears by codifying that the OPCW will:

> . . . [C]onduct its verification activities provided for under this
> Convention in the least intrusive manner possible consistent with
> the timely and efficient accomplishment of their objectives. It
> shall request only the information and data necessary to fulfill its
> responsibilities under this Convention. It shall take every precau-
> tion to protect the confidentiality of information on civil and
> military activities and facilities coming to its knowledge in the
> implementation of this Convention. . . .[22]

There is also a "Confidentiality Annex" that details how scientific or trade secrets should be protected. However, even among those states that have signed and ratified the CWC, considerable concern remains over the protection of proprietary information, as well as the fear that some countries might abuse the inspection process by harassing one another.

MONITORING AGENTS AND THEIR PRECURSORS

In order to comply with the provisions of the CWC, verification of this treaty entails the monitoring of CW agents and their precursors. Declarable facilities are defined in terms of the types of chemicals they produce, both in terms of risk to the CWC as well as the quantities manufactured.

Monitoring of CW Agents (Schedule 1)

Because Schedule 1 chemicals consist of CW agents and their immediate precursors, governments must notify the OPCW what CW agents are being produced, the amounts being made, and the purpose(s) for doing so, as only peaceful and defensive reasons are sanctioned. Due to the tightly controlled nature of production of such substances, domestic governments should have little trouble accounting for them. At the same time, it is hoped, inspections can rather quickly ascertain—again using sensitive instrumentation as well as the inspectors' expertise—the presence of illicit CW agents.

Monitoring of CW Precursors (Schedule 2)

In addition to BZ, Amiton, and PFIB, Schedule 2 of the CWC lists those chemical compounds that in combination with other substances could be used to produce CW agents or their precursors. For example, pinacolyl alcohol, a vital part of the nerve agent soman (GD), is a Schedule 2 compound. Countries that produce such substances, even for civilian industry, must declare the location, purpose, and quantities to the OPCW. Inspectors dispatched under the authority of the OPCW will then visit individual production facilities to check the accuracy of the declarations.

Monitoring of Commercial Chemicals (Schedule 3)

These materials are produced in large quantities for civilian industry, but also have potential military application. In the cases of chlorine and phosgene, a thorough, company-by-company accounting of each shipment's destination would be far to cumbersome and probably of little benefit in any case. Instead, countries that are State Parties submit reports that document amounts exported or imported of each material, and the destination or source, respectively. A country that exports 10,000 tons of chlorine a year, for example, will report as such, indicating also how much was imported or used internally. Wide discrepancies will alert the OPCW, which can decide whether further investigation is necessary.

Proliferation Signatures

While no single clue would necessarily be definitive evidence of CW agent production, taken in their aggregate, signs can point to processes unique to chemical weapons manufacture. Member nations and investigators thus look for "signatures" that suggest unsanctioned activity. For example, during the later stages of nerve agent manufacture—especially in the cases of DFP, sarin, and soman—non-corrosive reactor vessels and pipes, made of either glass or Teflon, are preferable. Similarly, special metallic formulations used in process equipment, which are acid-resistant and useful for nerve-agent synthesis (such as certain nickel or tantalum alloys), could indicate CW production. Pumps or valves used for hazardous or corrosive materials or high-volume air-filtration systems would also make inspectors suspicious, as would the presence of fluorine, phosphorus, or sulfur-assaying equipment, inert-gas generation equipment, and double-walled piping. In addition, redundant detection and alarm systems for toxic chemicals, glove boxes or automated filling rooms, and hazardous-chemical disposal and treatment equipment are also red flags.

Needless to say, such inspections can become both highly technical and extremely ambiguous. Because of recent attention to worker safety as well as environmental matters, many modern industries now employ a wide array of safety measures. Years ago, the mere presence of such protective equipment and process-safe designs would have provided a tell-tale signature of a CW facility, but this is no longer the case. So, the investigators have to take a broader view of a facility's characteristics, weighing the significance of details—a sophisticated air-filtration system, or excessive security features (armed guard posts, high fences), for example—that in other instances would not warrant a closer look. (In satellite photographs of a Libyan factory in the 1980s, the presence of anti-aircraft batteries around what was supposed to be a fertilizer plant was taken as an indication of something amiss.)

To produce a substantial chemical arsenal, facility and logistical infrastructure capable of handling tons of chemicals and reactant products as well as resulting wastes is needed. It is estimated that for the daily production of 10 tons of sarin nerve agent, one would generate about 40 truckloads of waste.[23] Even a facility built underground, a costly enterprise, would have to deal with by-products in some fashion. In some cases, the examination of these effluents can be done in a remote location and far from the facility. The detection of certain degradation products can indicate the presence of precursors, and even the chemical agents themselves. The phosphorous–carbon bond found in most nerve agents, for example, is very stable and remains intact for a relatively long time in soil.

Aside from the presence of the actual CW agent itself, related by-products, and the technology required to produce them, there are other clues that may indicate the production of chemical weapons. Unusual environmental changes or damage can be an indicator, although such clues are hardly definitive. The death of plants and browning of leaves surrounding the Aum Shinrikyo cult's chemical facility was noticed by neighbors, but it took spectrographic analysis to prove that sarin was being produced. In addition to environmental damage, large amounts of certain biological activities, such as accelerated eutrophication, or the excessive blooming of algae, are often an indicator of high phosphorous levels, which in turn may signal the presence of chemical weapons manufacture. But, as with other indicators, the mere presence of eutrophication is not definitive, since it may also be triggered by the presence of phosphates from detergents and other sources.

Challenge Inspections

The challenge inspection regime, a critical part of the verification and compliance process contained in the CWC, permits a State Party to request, on short notice, an inspection of *any* facility, private or government-owned, declared or undeclared, on the territory of another State Party or under its jurisdiction and control. If a state that is party to the CWC believes another is in violation, that state will notify the Director-General of the OPCW. An Executive Council of 41 member-states evaluates the substance of the charge, and if it is seen as "frivolous or abusive,"[24] a three-quarter majority vote can block a challenge inspection. Otherwise, the inspection will proceed. As it stood in 2001, and despite the Convention having been in force since April 1997, no country has called for a challenge inspection.

As its name implies, challenge inspections are conducted with limited notification, usually 12 hours, before the inspection team arrives. The greatly abbreviated notice prevents a potential violator from being able to remove equipment, munitions, or other evidence that would point to chemical weapons manufacture. After the inspection team arrives, the facility management must give access to the inspectors within at least 120 hours following notification.

In the event of alleged use of chemical weapons on a state's territory by another, an inspection can also be requested. An investigation of this type would attempt to verify the claim, ascertaining what, where, and when chemical weapons were in fact used. But apart from earning the collective opprobrium from the international community and losing the benefits of being a state party to the CWC, there are no clearly stated repercussions in terms of punitive measures.

Managed Access

Working in an extremely competitive field, chemical industry have been concerned that intrusive inspections could lead to the disclosure of their closely guarded trade secrets. However, protecting confidentiality can be achieved in a number of simple ways. Files, papers, computer screens, and process control panels can be shrouded during an inspection, provided that the facility can demonstrate that these areas are non-CWC related. Furthermore, proprietary process techniques that do not involve CW agents do not need to be described to the inspectors in great detail.

Some member states have suggested an alternative approach to the intrusive inspection of chemical facilities, thus avoiding sticky confidentiality issues entirely. With modern analytical instruments, the detection of extremely small amounts of suspicious compounds is possible. By first sampling the facility's wastewater for typical signs of CW agent manufacture, inspectors can readily ascertain possible violations without an on-site inspection of a suspected chemical weapons plant. However, even if no compounds are detected, a "clean bill of health" can not be certified. Furthermore, even a clandestine production may have a fully capable waste treatment facility on site. Therefore, while such monitoring methods are useful, full compliance with CWC provisions is best assured by first hand inspection of facilities and equipment.

DESTRUCTION OF CHEMICAL WEAPONS

Under the original US Department of Defense Authorization Act of 1986 and its amendments, the United States must destroy its stockpile of chemical weapons by 2004. Following the US Senate ratification of the CWC (1997), and despite some technical challenges, this destruction should take place within the international schedule.

Earlier disposal techniques were often too expedient and done without much thought of the ecological consequences. For example, following World War II, the Allied forces dumped 250,000 tons of Germany's chemical munitions, most of it going straight into the Baltic Sea. Japan's remaining chemical stocks were dispensed in similar fashion off its coast in 1945. In the 1960s, the US Chemical Corps disposed of leaking M55 nerve agent rockets by firing them into the desert where, upon exploding, their contents would dissipate over time.[25] Before the end of World War II, approximately 69,000 artillery shells filled with tabun (GA) nerve agent were dumped by the German government in relatively shallow water (20–30 meters) between two Danish islands. Germany recovered these munitions in 1960 so that they could be more adequately disposed of in the Bay of Biscay.

In the United States today, the government is faced with the challenge of disposing of about 30,000 tons of CW agents. Clearly, dumping them at sea or using other hazardous methods is illegal and, considering the current climate of environmental awareness, is out of the question in any case. In the US, the destruction of CW agents and munitions involves three stages. First, the agent must be seperated from the munitions or other container by automated handling equipment. Second, the liquid chemical agent is collected in storage tanks from which the agent is fed into a high-temperature incinerator and burned at a temperature of 2700 degress Fahrenheit. The "off gasses" generated by the high-temperature combustion process are reprocessed, and the agent is neutralized. Third, the chemical weapons are fully decontaminated and the rockets or bombs rendered inoperative, and finally disposed.

PRIMA CORD IN GERMAN ROCKET BASES- HN₃ FILLED

Chemical warfare agents such as nitrogen mustard were stockpiled by Germany, but never used in World War II. These 150-mm nitrogen mustard (HN-3) rockets are shown being prepared for destruction by the Allies. (Courtesy of Soldier Biological and Chemical Command, Historical Research and Response Team, Aberdeen Proving Ground, MD.)

Not in My Backyard

Communities near chemical weapon storage depots in the United States are understandably worried about accidental spills or releases. M55 rockets containing large quantities of sarin (SB) are of particular concern. Some of this ordnance has been known to leak, and there is a very small risk of explosion from propellants. Extra safety precautions have been instituted to ensure safe incineration and to limit the amounts of effluent released into the environment. Despite the risks inherent in the destruction process, the dangers in allowing the weapons to rust and leak have been determined to be greater than carrying out the disposal.

Due to public protest and controversy surrounding incineration as a method for disposal, the US has researched alternative technologies. After considerable study and recommendations made by the National Research Council Panel on Alternative Chemical Disposal Technologies, Aberdeen, Maryland, and Newport, Indiana, were chosen as chemical neutralization sites. Both sites held CW agents in bulk form, making such decontamination methods more practical. Aberdeen only stores mustard agent, which means that other novel treatments using microbes could also be used after neutralization.

The Destruction and Conversion of Facilities

Because of the unitary purpose of CW agent production facilities and concerns about contamination, most US chemical weapon manufacturing sites are no longer maintained. Demolishing these facilities will not pose much of an economic burden other than the cost of destruction and clean-up.

In the former Soviet Union, however, the situation is rather different. Many of the civilian chemical industries in Russia are tied into plants that used to manufacture CW agents. Plants that produced Lewisite and others that were heavily contaminated by nerve agents have already been destroyed. But Russian politicians have since balked at the wholesale destruction of other CW agent facilities, claiming that they can be legally and safely converted to civilian industrial use. For example, some of the same facilities that manufactured nerve agent precursors could be converted to produce valuable chemical products for the consumer industry. However, the CWC needs to have the resources to monitor the converted facilities to ensure they do not resume CW agent production. Some American businesses have expressed reserved interest in forming joint ventures, but understandably would like to know more about the conditions at Russian facilities before making large investments.[26]

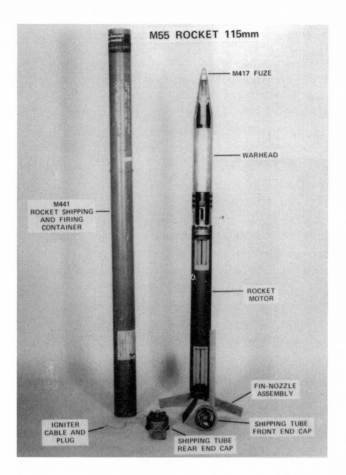

The M55 unitary 115-mm sarin or VX nerve agent rockets were produced for delivery from the US Multiple Launch Rocket System. However, a design flaw in the warhead caused leakage almost as soon as they were manufactured. (Courtesy of Soldier Biological and Chemical Command, Historical Research and Response Team, Aberdeen Proving Ground, MD.)

Brain Drain in the Former Soviet Union

While the CWC focuses on the movement of technologies and of compounds, they are equally concerned about the movement of knowledge.
Rogue nations, now euphemistically referred to by the US State Department
as "countries of concern," and well-funded terrorist groups can offer attractive
salaries and benefits to unemployed Russian chemists. As a consequence, and
despite the risk that converted chemical facilities in Russia could continue to
manufacture CW agents, it might be advantageous to keep such plants online.
Even if converted CW agent production facilities do not turn out to be prof-
itable, keeping the technicians, scientists, and engineers working at such sites
may help to prevent "brain drain." Although this would amount to a subsidy of
certain parts of the Russian chemical industry, it would help curb the spread
of chemical weapon manufacturing expertise.

Concern over brain drain—and that former Soviet scientists would be
recruited by rogue regimes such as Libya and North Korea—led to the formation of the International Science and Technology Center (ISTC) in Moscow.
Funded by a consortium of European countries, the United States, and Japan,
ISTC began funding projects to employ former nuclear, biological, and chemical weapons engineers in peaceful research activities. ISTC has employed over
17,000 former Soviet scientists, engineers, and technicians. By 1997, over $145
million had been spent for ISTC-supported research projects, with the
Kurchatov atomic research organization and the BW apparatus known as
Vector receiving the lion's share of funding.

According to a review by the US National Research Council, the program
has been successful in recruiting weapons scientists and keeping them
employed in non-military research. Between the ISTC initiative and corporate R&D investments from the West, by 1997 over $500 million in coopera-
tive research funds had been raised that can continue the hiring of former
Soviet weapons experts.[27] The Soros International Science Foundation and
the Russian-American Biomedical Research Foundation have also contributed to this endeavor. However, it is difficult to gauge how effective the
ISTC program (and others) have been or will be in preventing the proliferation of expertise in chemical weapons manufacture.

PART THREE

Biological Agents

CHAPTER 7

Basic Concepts

If one can define chemical warfare as the utilization of toxicology toward malevolent ends, then biological warfare (BW) is truly, as Brigadier General William M. Creasy of the US Army Chemical Corps put it in 1952, "public health . . . in reverse."[1] Whether it is targeted against humans, animals, or plants, the use of biological weapons deliberately sets infectious disease into motion, which is why biological weapons are so universally reviled, even more than chemical weapons. To many if not most people, the unleashing of a disease-causing bacterium, virus, or other biological agent on humanity is the epitome of evil.

Biological weapons dispense, project, or disseminate disease-causing and poisonous agents for use in war or terrorism. These may take the form of missile warheads, aerial bombs, or line-spray devices, but also include poisoned projectiles such as flechettes. Because they rely on making large numbers of people sick and do not necessarily destroy property, biological weapons might be more accurately termed "mass casualty weapons"[2] rather than weapons of mass destruction (WMD).

Biological weapons primarily function by inhalation, by infecting open wounds, or by contaminating food and water. Most BW agents are effective when delivered as tiny particles in aerosols that can be inhaled deep into the lungs. In order to find their way to the alveoli (the tiny air sacs in the lungs), particles of about 1 to 5 microns in size are optimal. Droplets this small reach the lower parts of the lung, breaching the blood–gas barrier to start an infection.

While thousands of microbial pathogens and toxins occur in nature, only some 160 of these have been recognized as having the capacity to harm humans; only about 30 of these are considered likely biological warfare agents. These agents have been tested and developed for use as BW agents, some proving to be more effective than others.

Biological agents considered suitable for weaponization share one or more of the following features (few of us have any innate immunity to BW agents and their associated diseases):[3]

- The ability to infect reliably in small doses
- High virulence, or capacity to cause acute illness resulting in death or incapacitation
- The ability to remain potent during production, storage, and handling
- Short incubation period between infection and onset of symptoms
- Resistance to medical treatment[4]
- Suitability for economic production in militarily significant quantities
- Ease of dissemination and ability to survive environmental stresses
- The attacking force employing BW is protected against the agent.

As in chemical warfare, the goal of BW is to incapacitate and not necessarily kill enemy personnel. For this reason, many agents or pathogens are suitable in BW because of their high infectivity rather than lethality. Q-Fever, for example, is almost never fatal, but a single organism of the causative agent (*Coxiella burnetii*) is sufficient to cause an infection through the lungs. (Additionally, Q-Fever bacteria form spores, which make it more stable during delivery). The Venezuelan equine encephalitis virus causes a disease that is rarely fatal, but was developed by both the US and Soviet BW programs because of its high infectivity.

Some very contagious pathogens can have a boomerang effect on the user, but this does not rule them out. Scientists in the Soviet Union, for example, experimented with the variola virus that causes smallpox, the extremely contagious disease that is about 30-percent fatal. Smallpox was eradicated twenty years ago and has not been seen in humans since a Somalian case in October 1977 and two laboratory infections in the United Kingdom a year later. Today, nearly everyone in the United States is susceptible to variola infection because smallpox vaccinations have not been required for civilians since the early 1970s and were discontinued for US military personnel in 1989. Ken Alibek, a former deputy director of the Soviet BW research and development effort, reported that the USSR at one time targeted the US with intercontinental ballistic missiles armed with warheads containing smallpox virus in a liquid form. According to Alibek, the Soviets reasoned they would have been safe from this contagious biological agent because of the geographic distance between the two nations.[5]

Modern bioengineering techniques can be used to enhance existing biological agents and make them ideal biological weapons. For example, antibiotic-resistant bacteria can be cultured in the laboratory, an activity carried out in the Soviet Union in the late 1980s. Gene-splicing techniques may make agents more virulent or better able to survive and remain potent until they reach their target hosts. On the other hand, since wholly new diseases would be extremely difficult to create, biological weapon proliferators (terrorist groups as well as states) are more likely to work with biological agents that have previously been found suitable for weaponization.

Although biological warfare has a long history (see Chapter 8), the actual use of biological weapons in modern times has been very rare. Nonetheless, because the proliferation of BW-related technology has increased dramatically over the last few decades, the threat of a biological weapon being used in war or terrorism is very real. At least 17 countries are known to have or are suspected of having BW programs, while as many as 100 have the requisite expertise, equipment, and infrastructure to create their own.[6] Developing countries, in particular, may view biological weaponry as a cheaper means to counter nuclear-armed neighbors or regional superpowers.

By their very nature, BW agents have a delayed effect, a characteristic that makes them disadvantageous for some potential users but highly attractive for others. For the purposes of terrorism this feature could be useful: the lag time between the dissemination of a BW agent and the appearance of symptoms of infection in the targeted population could allow users to get away and cover their tracks.

Biological weapons should be considered strategic weapons because they have the potential to inflict massive casualties over a wide area. Delivering a kilogram of anthrax—in an effectively distributed form under optimal meteorological conditions—could kill hundreds of thousands of people in a metropolitan area. Decades-old field experiments using animals as targets underscore the possible catastrophe that could occur from a BW agent attack.

During the 1960s, the discontinued American offensive BW program, tested in the Pacific Ocean, demonstrated the viability of biological agents as strategic weapons. Before the end of the BW program in 1969 and 1970, US scientists and military technicians tested aerosolized lethal and nonlethal agent powders on live animals. A series of experiments conducted over the Pacific Ocean downwind of Johnston Atoll (a thousand miles southwest of Hawaii) showed how effective biological weapons could be against unprotected populations. In 1968 near Johnston Atoll, monitored by Navy personnel wearing protective suits, Phantom jets released in powdered form an unidentified BW agent[7] from pods under their wings. Floating miles downwind were barges carrying hundreds of rhesus monkeys. About half of the monkeys died within days of the attack. From these tests, US scientists concluded that a BW attack conducted against a large city could kill as much as 50 percent of the population. In 1993, General Colin Powell, former Chairman of the Joint Chiefs of Staff and now US Secretary of State, testified in a Congressional hearing: "The one that scares me to death, perhaps even more so than tactical nuclear weapons, and the one we have the least capability against is biological weapons."[8] Since the end of the Cold War, the perceived threat of biological weapons is increasingly widespread. On the other hand, experts such as Elisa Harris, who formerly served on the US National Security Council, believe that only the "perception of the threat, not the reality of the threat" has changed.[9] On an individual basis, the risk of encountering BW in one's lifetime may be quite low; nevertheless, in the wake of the events of September 11, 2001, the potential consequences of even one well-delivered attack using biological weapons are too great to ignore.

Because of the basic principles of biological agent delivery and their characteristics, biological weapons and their agents are best understood apart from chemical weapons. To help understand the contrast between chemical and biological weapons, recall the behavior and properties of CW agents. A relatively large liquid droplet of VX nerve agent (say, 20 microns or more in diameter) is much less likely to be inhaled very deeply into the lungs but still can easily be absorbed through the skin or upper respiratory tract. Sarin, a relatively volatile nerve agent, need not be disseminated as an aerosol because its amorphous vapor is quite deadly. Finally, liquid mustard can be delivered en masse, covering ground soil or objects, and presenting both a vapor and contact hazard to troops that may walk over the contaminated area. By contrast, save for only one BW agent (the toxin T2, derived from *Fusarium* molds), the effective use of BW agents is primarily achieved though the means of inhaled aerosols.

BIOLOGICAL WARFARE AGENTS

Biological weapons employ living organisms or biologically produced toxins to injure or kill humans, animals, or crops. Biological warfare agents include pathogens (microbes that cause disease) such as bacteria, viruses, and fungi. BW toxins are those produced by bacteria, fungi, or other living organisms.

In the case of bacterial agents used as weapons, these microbes often are the source of toxic substances that injure or kill. *Bacillus anthracis*, the bacteria responsible for anthrax, is an example of a BW agent pathogen that produces a lethal toxin complex in the body. Viruses, on the other hand, are strictly *obligate* parasites, that is, they cannot replicate without direct assistance from living cells.[10] Viruses can destroy cells, wreaking havoc as the host puts up a defense. Viruses that have been mentioned in a BW context or actually weaponized include smallpox, yellow fever, Marburg (a close cousin to Ebola) virus, and foot-and-mouth disease virus for use against livestock animals. The most important of these viruses is smallpox, which would be an unmitigated disaster if it were to reemerge.

Certain pathogens such as bacteria, and possibly viruses, can be cultured or manipulated in certain ways to increase their virulence.[11] In the 1980s, Soviet BW scientists also developed strains of bacteria that would be resistant to certain antibiotics.[12] Genetically altering bacteria, for example, might include emphasizing traits to increase their virulence and overall suitability for weaponization. Ken Alibek reports that Russia engineered a so-called chimera virus. Like its namesake, the mythical beast made of different creatures, this was a smallpox viron combined with the Venezuelan Equine Encephalitis virus.[13]

While some biological toxins can be synthesized artificially, the practical method for producing these poisonous substances is to extract them from living organisms. For example, botulinum toxin is the active poison in the bacterium *Clostridium botulinum*. Trichothecene or T2 toxin is derived from species of *Fusarium* mold, and is unique among BW agents for its ability to cause pain and injury on contact with unprotected skin.

Consisting of a large protein structure, botulinum toxin would be impractical to manufacture in large quantities without the assistance of microbes. However, another BW agent, saxitoxin—found in certain algae and shellfish— is relatively simple in its chemical structure and has been artificially synthesized in the laboratory. BW toxins could also be manufactured in large quantities by genetically altering more common bacteria found in humans and animals (such as *E. coli*). This toxin could then be extracted by precipitating the protein from the cultured material, or the otherwise innocuous organism could be used as a BW agent. (Researchers in India, for example, successfully altered *E. coli* to manufacture the so-called Lethal Factor toxin found in *Bacillus anthracis*). BW agents offer some distinct advantages over chemicals. Under the right conditions, BW agents can produce a larger "footprint," that is, cover a wider area than an equal amount of chemical agent. According to one authoritative source, "Studies using computer models have shown that clouds of hardy organisms such as anthrax spores can be infectious more than 200 kilometers [125 miles] from the source of the aerosol."[14] Very small quantities of a microbial pathogen can cause tremendous harm. A lethal dose of anthrax bacteria in humans is variously estimated at 8000 spores. This may seem like a high number, but it must be kept in mind that each spore is extremely small, measuring about 1 micron in diameter, while a cubic foot of normal breathable air can easily contain thousands of microbes. It would not require many breaths of air laden with anthrax spores to achieve an infectious dose, which would mean death without prompt treatment.

Thus, on a pound-for-pound basis, biological agents are considerably more potent than chemical weaponry, and the costs of BW agent development are also lower. Especially when compared with the manufacture of nuclear or even some chemical weapons, the technical threshold in developing microorganisms for a biological weapon is low. On the other hand, engineering devices to deliver BW agents is more difficult.

THE NATURE OF INFECTIOUS DISEASE

Unlike chemical warfare, which is based on compounds known since ancient times, the knowledge of microbial disease processes is a relatively new science. Not until the late 1800s was it demonstrated that microbial agents or germs could be the cause of infection.

The Germ Theory of Disease

In the late 1500s, the brilliant Italian physician Girolamo Fracastoro theorized that the disease of syphilis was an infection "that passes from one thing to another." According to this view, there were three main ways that the contagion could be spread: 1) by close contact, 2) by fomites (contaminated objects), and 3) from a distance.

In a similar vein, a seventeenth-century German Jesuit, Athanasius Kircher (1602–1680), reported on the plague outbreak in Rome in 1656. A contemporary of Antony van Leeuwenhoek, the Dutch pioneer of microscopy, Kircher may have been the first to use a microscope for the purpose of examining disease. Looking at specimens from plague victims in Rome, Kircher wrote in his *Scrutinium pestis* that he saw "small worms," although it is doubtful he actually saw plague bacteria.[15] Still, this seemed to add further evidence to support the contagion theory even at its earliest stages.

Despite the work of Fracastoro, Kircher, and others, two centuries passed before microorganisms were acknowledged as having an essential role in illness. Up to the late nineteenth century the main etiology of disease was variously ascribed to various miasmas, "noxious effluvia,"[16] or "corrupt vapors." In this older conception, the cause of disease had much more to do with astrology or evil machinations than, say, the absence of proper hygiene. Even by the mid-nineteenth century, the importance of something so simple as doctors washing their hands to reduce childbed (purpeural) fever[17] was openly ridiculed.

The Advent of Modern Microbiology

As we have seen, during the nineteenth century great advances in chemistry came about with the introduction of anesthetics and synthetic dyes. These developments also contributed to a new understanding of infectious disease. Improvements in the light microscope, along with the development of synthetic dyes and staining methods, made it easier to identify and differentiate bacteria from other cells, and even to clearly see smaller organs (organelles) within microbes.[18]

Louis Pasteur, who began his career as a chemist, was the first to prove that germs cause disease. Addressing the French Academy of Sciences in 1878, Pasteur declared:

> When, as the result of my first communications on the fermentations in 1857–1858, it appeared that the ferments, properly so-called, are living beings, that the germs of microscopic organisms abound in the surface of all objects, in the air and in water; that the theory of spontaneous generation is chimerical; that wines, beer, vinegar, the blood, urine and all the fluids of the body undergo none of their usual changes in pure air, both Medicine and Surgery received fresh stimulation.[19]

Paul Ehrlich, the founder of modern immunology, seizing upon how coloring agents attach themselves to textiles, discovered that certain chemicals bind more aggressively to some cells than others, and that this could be used to selectively kill microorganisms. Building upon this, Ehrlich presaged modern antibiotics by developing the first effective agent against syphilis, the arsenical Salvarsan, in 1890. Meanwhile, in the 1870s, another German microbiologist, Robert Koch, studied a devastating disease known as the "Siberian plague," which is commonly known today as anthrax. In 1876, Koch isolated *Bacillus anthracis* and showed it to be the causative agent of anthrax in horses. Koch probably was the first to demonstrate that anthrax bacteria could form hardy spores to allow their survival in soil. Using anthrax as a model, Koch further established the rules for determining the pathogenesis of microbial disease. Now known as Koch's postulates, these basic rules state that, in order to determine a microbe is the cause of disease,

- the organism must be found in the diseased animal but not in the healthy ones;
- the organism must be isolated from diseased animals and grown in a pure culture;
- the organism isolated in a pure culture must initiate and reproduce the disease when reinoculated into susceptible animals;
- the organism should be reisolated from the experimentally infected animals.[20]

The Airborne Origin of Infectious Disease

In the 1860s, Pasteur showed that living microbes can be found in the air. In 1883, Koch took air samples in both Germany and England, finding 58 times more microbes in London air than in Berlin air.

It was another matter to show whether or not microbes in the air, suspended in aerosolized particles, could be responsible for the transmission of disease. While there may have been strong suspicions that unseen particles floating in the air could cause disease, it remained a matter for dispute. Not until biological weapons research during World War II were lingering doubts cleared away.

DIFFERENTIATING AMONG PATHOGENS

Bacteria

In 1884, experimenting with new methods of staining microbes to better examine them through the microscope, a Danish medical researcher, Hans Christian Gram, found he was able to stain some bacteria with his procedure but not others. Gram went to his grave not knowing that his "failure" would add significantly to the framework of microbiological science. As it happens, bacteria that accepted Gram's stain are taxonomically distinct from bacteria that did not. Today, these are classified respectively as Gram-positive and Gram-negative.

The principal difference between Gram-positive and Gram-negative bacteria is in the structure of their cell walls. Gram-positive bacteria have a thick cell wall layer of a protein-sugar complex called peptidoglycan. Gram-negative bacteria possess a very thin layer of peptidoglycan, but their cell wall is more complex. Built into the walls of Gram-negative bacteria feature is a so-called endotoxin, lipopolysaccharide (LPS), one of the deadliest substances known. Some Gram-negative bacteria are noted for their ability to cause toxic shock, bringing about a dramatic lowering of blood pressure that frequently results in death. On the other hand, Gram-positive bacteria produce so-called exotoxins which are also deadly but are especially prone to generate fever.

The Rickettsiae

Another category of bacteria that plays a significant role in disease are the so-called rickettsial organisms, named after Harold T. Ricketts, who in 1907 first demonstrated the transmission of Rocky Mountain Spotted Fever (caused by a bacterium now named *Rickettsia rickettsii*) from a human to a guinea pig. Rocky Mountain Spotted Fever was determined by Ricketts to be carried by arthropods, i.e., insects, spiders, and so on, in this case by ticks. While given a separate family designation within the biological order, rickettsial organisms are considered Gram-negative bacilli that for the most part become parasitic at a stage in their life cycle.

Long the scourge of armies, the best-known rickettsial-caused disease is typhus, responsible for more deaths from disease than any other bacteria. Another Gram-negative member of the family Rickettsiae is *Bartonella* (*Rochilamaea*) *quintana*, a common cause of trench fever during World War I. Q-Fever is caused by a separate member of the Rickettsiae family and is unique in that it forms a hardy spore. In 1937, two researchers named Cox and Burnet independently described the agent of Q-Fever, which was named *Coxiella burnetii*. The so-called Balkan grippe, which during World War II infected many German soldiers, was caused by *Coxiella burnetii*.

Typhus is caused by the rickettsial *Rickettsia prowazekii* and most often spread by another arthropod, the body louse. The Latin name was given in honor of Ricketts and another pioneering researcher, Stanislas von Prowazek. Both scientists eventually died from typhus as a result of their research.

Ranked alongside plague and malaria, *Rickettsia prowazekii* and its arthropodic vectors (carriers) are responsible for much of the death and suffering from disease throughout history. Following Napoleon's invasion of Russia in 1812, France lost up to 80,000 men from disease, mostly epidemic typhus. Tight living quarters and poor hygiene contributed to infestation by lice and the spread of "spotted fever," i.e., typhus. A century later, in the formative years of the Soviet Union, typhus was rampant. From 1918 to 1921, as many as 30 million people were infected, and 10 percent of these died. This experience helped convince Red Army commanders to pursue biological weapons development.

Viruses

The term "virus" comes from the Latin word for poison. Peter Medawar, a biologist who was awarded the Nobel Prize in 1960 for his contributions to medicine, defined a virus as a "piece of nucleic acid surrounded by bad news."[21] Viruses can multiply only by means of a parasitic relationship, be it harmful or innocuous, with bacteria, plant, or animal cells. It is viruses that cause influenza (flu) and the common cold.

Viruses are tiny by comparison to bacteria. Light microscopes cannot resolve images smaller than 0.2 microns (or 200 nanometers) in size, making nearly all viruses undetectable except by advanced electron microscopes (the smallpox virus, at about 230 × 300 nanometers, is a notable exception). As a result, it took much longer to determine the etiological role of viruses in disease. Though Pasteur was unable to identify viruses as such, his use of laboratory animals revolutionized the study of all microbial pathogens, including viruses; he also made an important study of rabies, which is now known to be caused by a virus.

In 1892, a Russian botanist by the name of Dimitrii Ivanovski experimented with diseased tobacco plants. After passing liquid from these plants through fine porcelain filters, a process that normally removed bacteria, Ivanovski found that healthy plants developed tobacco mosaic disease when treated with this fluid. Being a cautious and responsible scientist, Ivanovski could only surmise that his filters were of poor quality and disease-causing bacteria simply had gotten through.

In 1898, after more or less duplicating Ivanovsky's experiments with tobacco plants, Martimus Willem Beijerinck, a Dutch scientist, wrote in a paper that there must be a causative agent for tobacco mosaic disease—not a bacteria but a "virus" that could not be filtered out with his laboratory apparatus.[22] In the same year, Friedrich Löffler and Paul Frosch in Germany discovered that foot-and-mouth disease—an illness of pigs and cattle—could be passed from one animal to another using infectious fluid put through porcelain filters. Finally, in 1900 Walter Reed and James Carroll showed that yellow fever-contaminated material remained infectious after filtering.

Bioaerosols

The use of the word "aerosol"—referring to the suspension of liquid droplets or particles in the air—goes back to World War I, but the role played by airborne, micron-sized particles bearing pathogenic organisms in causing infection was not well known until the mid-1930s.[23] But it was already established that the smaller the particle the slower was the rate at which it settled; likewise, it was found that the smaller the particle the longer it would remain aloft.

It has been estimated that microscopic particles—too small to be seen by the naked eye but large enough to be seen by light microscopy—exist at a concentration of about 1 million per liter of air, possibly ten times this amount in industrialized areas. The most commonplace aerosolized particles range from about 0.1 to 10 microns—again, too small to be seen by the naked eye. (Human vision cannot distinguish objects smaller than 30 microns in size. Fog, for example, is caused by a dense collection of water droplet/particles, ranging from about 4 to 40 microns in diameter.)

If the airborne particles are small enough, disease-causing microbes gain access to the lower stages of the lungs where bronchial tubules become progressively narrow. Infection is more likely to ensue when the agent particles reach further down in the lungs to the alveolar spaces. It is in the tiny alveoli where gas exchange takes place and where pathogens can enter the host. Particles that range in size from 0.5 up to 3 microns in diameter can find their way to the alveoli. (Particles below 0.5 microns actually deposit at a lower rate, usually being exhaled before making any impact on tissue.) Larger particles, from 3 to 5 microns in size, become blocked by upper respiratory barriers. Particles larger than 5 microns will more likely be trapped and ejected by the upper respiratory tract. Most microbe-laden particles that cannot penetrate below the upper respiratory tract will have little consequence.

This helps to understand the transmission of certain diseases by air. Valley Fever, for example, is a sometimes very serious disease caused by the fungal agent *Coccidioides immitis* (CI). Infection starts with the inhaling of airborne particles, most of these averaging less than 10 microns in diameter.[24] Over 80 percent of people in certain endemic areas of the western United States test positive for having been exposed to CI,[25] a clear demonstration of how infectious particles can spread by aerosol.

A classic case of infection by aerosol is the disease known as psittacosis or ornithosis. Recognized since the late 1800s, it was only in the 1930s that global attention focused on this particular illness, prevalent both in animals and humans. In one scenario, a pet bird such as a parrot develops an infection by *Chlamydia psittaci* bacteria. During the course of its infection, when the parrot ruffles its feathers, dried feces or other matter are stirred up, creating a source of infectious particles. These remain airborne long enough for a member of the pet owner's household breathe them in and contract the disease. In another finding, between 1948 and 1953, workers at a turkey processing plant employed in slaughtering and feather picking contracted psittacosis, while those who graded and packed turkeys did not become infected. In 1950, the US government conducted a BW simulation study, employing spores of a harmless bacterium called *Bacillus subtilis*, otherwise known as *Bacillus globigii* (BG). Because BG can form spores, these bacteria are capable of mimicking how anthrax-causing bacteria (*Bacillus anthracis*) might behave as an aerosol. During this particular experiment, the simulant bacteria was suspended in water at about 7 billion bacteria per milliliter of water (about 1 gram in terms of weight). Sprayers were used to create a cloud of particles 1 to 5 microns in size. While moving along a 2-mile course at a distance of 2 miles from land, approximately 130 gallons of this suspension was released as an aerosol from the stern of a Navy ship. After about half an hour, about 100 square miles was found to have been covered by the aerosol release.[26]

A 1959 study conducted on Q-Fever found infections among people working around domesticated animals, further evidence of the effects of bioaerosols. In this case, about 75 people who lived downwind from a sheep- and goat-processing plant became ill, while people living outside this "foot- print"—aside from a few who frequented the corridor—did not.[27]

BIOLOGICAL WARFARE AGENTS

Biological warfare involves the use of several broad categories of agents. Two of these are infectious microbial pathogens—bacteria and viruses—which cause disease in living humans, animals, or plants. The third is known as bio- logical toxins, poisons extracted from biological sources previous to their use as weapons.

The various BW agents, pathogens, and toxins are grouped below by gen- eral category, then listed in terms of their perceived threat, starting with the deadliest in each category. It will be noted that some, including foot-and- mouth disease and rinderpest, affect only animals, but their potential for caus- ing widespread damage to developed agriculture makes them highly suitable as weapons.

The bacteria that cause anthrax and the botulinum toxin are among the most likely candidates for weaponization. (Whether or not botulinum toxin would be an effective weapon, however, is a matter of controversy.) Viruses such as Ebola, while extremely deadly (up to 90 percent mortality in some instances), are difficult to weaponize due to their fragile structures. Ebola kills its victims by destroying cells, especially those that line blood vessels, finally bursting them open after the virus has replicated itself. Although their BW scientists found weaponizing Ebola quite difficult, a successful effort was made in the former Soviet Union to weaponize Marburg virus, taxonomically a close relative to Ebola.

For the most part, biological toxins poison their victims by attacking the nervous system, blocking the transmission in nerves and paralyzing the mus- cles needed in order to breathe. Saxitoxin is an example of this kind. Found in certain marine organisms, it operates by means of direct action on the nervous system, and can kill quickly. Botulinum toxin, on the other hand, must enter nerve endings before causing nerve blockage, and it takes from 24 to 72 hours for serious symptoms appear. Ricin, another lethal toxin, destroys the ability of cells to manufacture proteins, but again takes as long as 72 hours to begin killing its victims.

The quantity of a BW agent required to cause disease once inside a host and the level of injury that results vary widely. The bacterium that causes tularemia, *Francisella tularensis*, can infect humans via inhalation of fewer than 50 microbes; depending on the strain, 30–60 percent of those infected who are untreated will die within 30 days. Humans can be infected by only a single organism of *Coxiella burnetii* (causative agent of Q-Fever). Symptoms of Q-Fever, while temporarily incapacitating, are relatively mild compared to other BW agents: a person infected with Q-Fever can develop a fever, chills, cough, headache, weakness, and chest pain within as little as one week. The disease lasts for about two weeks, but patients rarely become critically ill or die. Ricin toxin, when inhaled or ingested, can produce symptoms such as fever, nausea, and chest tightness within a day, followed by death in less than one week.

Bacteria

The classification of bacteria also includes the rickettsial organisms that normally require a host in order to multiply. Bacteria are single-cell, free-living organisms that reproduce by division. Some of the more common bacteria are those found in animals as well as humans, such as *E. coli*. Being unicellular organisms, bacteria appear in varying forms and sizes, including coccus (spherical), spiral, and bacillus (rod-shaped), and differ in other characteristics, notably whether they can adapt to hostile conditions by forming spores. Bacteria cause disease in humans by invading tissues and/or producing toxins. BW agents that fall into this category include those that cause anthrax, plague, and tularemia.

As we have seen, rickettsiae are a "degenerate" form of bacteria that for the most part require a living host to continue growth. The classic example of rickettsiae infection is epidemic typhus caused by *Rickettsia*. Epidemic typhus (not to be confused with typhoid fever) has been a major source of infection among soldiers during war. A rickettsial bacterium more suited to use in BW is *Coxiella burnetii*, the causative agent in Q-Fever, which adopts a hardy spore-like form (unlike other true rickettsial organisms).

Anthrax

The causative agent of anthrax among animals and humans is the spore-forming, rod-shaped *Bacillus anthracis*. As demonstrated on farms and in certain textile industries, infection by *Bacillus anthracis* can be contracted by inhaling spores given off by wool or other animal fibers, giving it the name of "wool-sorters' disease." In addition, skin infections can be contracted from handling animals or meat products ("industrial anthrax"). In cutaneous anthrax, a black scab forms over a pustule, from which the infection got its name (*anthrakitis*, the Greek word for "coal"). Least common but potentially life-threatening is contracting anthrax by eating infected meat. Anthrax continues to be endemic among animals in many regions of Africa and Asia.

Once inhaled, anthrax spores can find their way to the alveoli, from which they are carried by other cells to the lymph nodes near the lungs where they germinate. It was once believed that anthrax bacteria, after growing in the bloodstream, blocked blood vessels, eventually causing death. Later, it was found that a toxin complex manufactured by *Bacillus anthracis* is the fundamental mechanism for the deadly disease.

Most microbial pathogens are too delicate to survive an explosion, heat, excessive sunlight, or other conditions. If the majority of the microbes die or toxins become sufficiently denatured, or destroyed, before reaching the targeted population, they will probably fail to cause harm. But some bacterial agents such as anthrax have the ability to survive a wide range of environmental stresses in the "dormant" form of spores, that is, building thick walls around themselves that protect them from the elements. Anthrax spores become active again upon reaching a good growth environment, such as a living human body. The ability to form spores makes anthrax a highly suitable candidate for a biological weapon, and it was a centerpiece of both Soviet and US BW programs.

Spore formation also makes anthrax a difficult microbe to kill. Its persistence was demonstrated by British experiments carried out on Guinard Island in Scotland during World War II. Following the conclusion of tests conducted on Guinard between 1942 and 1943, viable anthrax spores remained for decades. So hardy were they that only disinfectants like formaldehyde were strong enough to adequately decontaminate the island's topsoil.

The incubation period of anthrax infection is from a day to a week in length, probably depending on the number of spores inhaled. Victims experience fever, fatigue, and general malaise, but deceptively show improvement after the first symptoms. This is followed by a sudden respiratory crisis, then shock and death within a day and a half. Left untreated, cutaneous anthrax can create systemic infection and is about 20 percent fatal,[28] while in ingested and inhaled forms the rates are approximately 50 percent and 100 percent.

Before the United States ended its program in 1969, it had been developed an anthrax weapon. Until recently, a dried preparation of anthrax was an important part of the Soviet Union's strategic BW arsenal, and Iraq has admitted to UNSCOM inspectors in 1995 that it too had weaponized anthrax.

Plague

Classified many years ago under the genus *Pasteurella*, *Yersinia pestis* is a non-spore-forming, Gram-negative bacterium typically found among rodents, especially in areas where plague is endemic. For humans, there are two significant forms of the disease: bubonic plague, transmitted by flea bites, and secondary pneumonic plague spread in the air from person to person.

In biological warfare, the primary plague threat would come from aerosol-borne bacteria. Less likely would be the approach tried by Japan during World

War II which was to spread plague bacteria by using fleas as vectors. During its occupation of China and in subsequent testing on human prisoners, Japan grew fleas, infected them from rats, and stuffed them into clay bombs. Untreated Bubonic plague has a mortality rate of about 50 percent, while untreated pneumonic plague is 100 percent fatal. (If diagnosed early, plague can be successfully treated with antibiotics.) Incubation for pneumonic plague is approximately 2 to 3 days, and 2 to 10 days for the bubonic form. Death is caused by respiratory and circulatory collapse from septic shock, similar to that caused by the endotoxin produced by other Gram-negative bacteria.

Unlike anthrax, plague bacteria do not form spores, therefore are much more susceptible to environmental stresses and usually die after several hours of exposure to sunlight. Nonetheless, compared to other bacteria that do not form spores, plague bacteria are hardy. (The Soviet Union weaponized it for their BW arsenals, but American scientists—during the heyday of the US BW program—were unable to master the technique of mass producing *Yersinia pestis*.)

Tularemia

> Tula tula-remia,
> Tula tula rye,
> I've dressed six hundred bunnies,
> I think I'm gonna die!
>
> —*Anonymous*[29]

As the verse indicates, a danger from skinning tularemia-infected rabbits (and other rodents) has long been recognized by hunters and people in the fur trade. Tularemia (also called "rabbit fever") is caused by the bacterium *Francisella tularensis*. Both the United States and the Soviet Union weaponized tularemia, finding it to be an incapacitating BW agent.

Humans acquire tularemia occasionally via insect vectors, such as the North American tick. In aerosolized form, typhoidal or pneumonic tularemia can be transmitted to an individual by as few as 10 to 50 organisms. Pneumonia following tularemia can kill up to 35 percent of infected individuals. However, secondary infection via breath, from one person in close proximity to another, is extremely rare. With antibiotic treatment, the mortality rate from all forms of tularemia is less than 1 percent.

Glanders

The bacillus *Burkholderia mallei* is the causative organism of glanders, which usually infects donkeys, mules, and horses, although a few human cases occur from the handling of infected animals. When suspended in aerosol form, however, *B. mallei* is highly infectious for laboratory workers, requiring stringent safety measures for containment. Laboratory animals (hamsters) have been known to acquire glanders infection from as few as 1 to 10 organisms.

There is strong evidence that during World War I Germany infected beasts of burden used by Tsarist troops, reducing the capacity for Russia's logistical supply of artillery to the Eastern Front. In the 1930s and 1940s, the Japanese in China also infected humans, including prisoners of war, with glanders. In the same period, the US conducted research into glanders but did not go as far as weaponizing the agent.

Inhalation by human beings of the glanders agent can lead to a full-blown, systemic blood infection (septicemia), severe pulmonary infection, and chronic inflammation of the skin and eyes. There is an incubation period from 10–14 days, and septicemia can lead to smallpox-like rashes on the skin. Even with antibiotic treatment, dissemination of *B. mallei* in an aerosol could result in high mortality among humans.

Q-Fever

The rickettsial Q-Fever was so named because the source of its pathology was elusive at first—hence "Q" for "query."[30] Because Q-Fever can be transmitted via inhaled aerosols, farmers and those who work in slaughterhouses are most prone to infection. These microbes can survive drying and can infect by a single organism, making Q-Fever an excellent BW agent candidate. Q-Fever was a BW agent in the Soviet arsenal until replaced by other agents, such as glanders, in 1989.[31]

Though its symptoms—fever and difficulty breathing—can be incapacitating, Q-Fever is not lethal, and most patients recover on their own. Large amounts of *Coxiella burnetii* rickettsiae can be cultured in chicken eggs, for example, producing up to 20 billion organisms per milliliter (about one gram).

Cholera

Epidemic cholera is a common cause of illness in countries with inadequate water treatment facilities. Cholera excretes a toxin (choleragen) consisting of a binding agent that attaches the microbe to the intestinal wall. The microbe then penetrates mucus layers, which produces the classic symptom of diarrhea. In the case of Asiatic cholera, untreated cases of infection have an approximate fatality rate of over 60 percent. Diarrhea, vomiting, and malaise appear after an incubation period from 12 hours to 3 days, symptoms caused by severe water loss and subsequent lack of body electrolytes.

The likeliest use of cholera as a weapon would be the contamination of a water supply or as a breathable aerosol. But this Gram-negative bacterium cannot survive acidic conditions, nor does it survive long in the absence of water. Given modern standards of hygiene interventions and the chlorination of water supplies, cholera is an unlikely modern BW threat. The bacterium does not form spores thus it does not survive harsh conditions, and many of those who are infected will have no symptoms, making its use as a weapon even less likely.

Viruses

Viruses cannot replicate by themselves. Like many of the rickettsial bacteria, viruses require a host cell in order to reproduce and are technically described as obligate intracellular parasites. But viruses are many times smaller than bacteria, the latter being about 1 micron in size on average. Smallpox is among the largest viruses known (about 0.3 microns in diameter) while the smallest known disease-causing virus in humans is the poliovirus (0.028 microns).

One theory concerning the origin of viruses suggests that they were formerly genetic material from cellular organisms that somehow exited from the cell, whereupon this genetic information associated itself with other protein-like groups, using living cells as hosts.

Viruses cause the majority of diseases in humans, ranging from the common cold to the extremely lethal Ebola. Other diseases include Venezuelan equine encephalitis and AIDS. The cause of Korean hemorrhagic fever, Hantavirus[32] (Hantaan or Bunyavirus) was named after the site where it was first identified, and a second group of the Hantavirus (the so-called *Sin Nombre* virus) was the source of a fatal outbreak in the southwest United States in 1993.[33] Hantaviruses and their cousins have also been considered—at one time or another—candidates for BW agent development by the major powers. But the largest BW threats include smallpox and the hemorrhagic fevers of the Marburg and the closely related Ebola viruses.

Smallpox

During the eradication of smallpox in the 1970s, it was seen as an unlikely threat for countries that engage in systematical periodic vaccination. The United States now considers smallpox a much higher threat for the following reasons: 1) the relative ease of culturing the virus; 2) a growing segment of the world's population that is not immune (smallpox vaccinations for most US citizens ended by 1971); 3) high infectivity through aerosol (primary and secondary) transmission; and 4) knowledge that the Soviet Union weaponized smallpox for strategic missile delivery during the Cold War. In the former Soviet Union, no less than 20 tons of smallpox were ready at any given time to be loaded into intercontinental ballistic missiles.

At present, the only known repositories of smallpox virus are in the United States and the former Soviet Union: at the Centers for Disease Control and Prevention in Atlanta, Georgia, and the Russian State Research Center of Virology and Biotechnology near Novosibirsk, in Siberia. Other nations such as North Korea and Iraq may also have viable stocks of *Variola major* stored at undeclared sites, possibly for use as biological weapons.

Smallpox can be transmitted by inhalation of the virus suspended in aerosols. After about a 12-day incubation period, infection from smallpox causes fever and headache. As the virus spreads to the skin it forms pus-filled vesicles across the body. Survivors usually are noticeably scarred for life. The mortality rate for immunized individuals is approximately 3 percent, while for non-immunized humans it increases to 30 percent.

Additional research carried out in Russia within the last decade seems to have involved the genetic manipulation of variola-type viruses (using monkey pox or other pox, rabbit, and monkey-pox viruses) as models to create an especially virulent form of smallpox. Iraq, for example, experimented with camel pox, possibly as a means to simulate the delivery of a smallpox weapon for testing.

The United States stopped vaccinating its military population in 1989 and civilians in the early 1980s, although an ambitious program to stockpile vaccines in case of a future BW attack was proposed in 2000. Instead of vaccinating its troops for smallpox, the US is counting on the availability of existing vaccine stocks since smallpox vaccine can be an effective prophylactic treatment if administered within one week of infection. In the future, some of the newer antiviral drugs could be used for the treatment of smallpox, although thus far chemotherapy in this area has not been promising.

Hemorrhagic Fever Viruses

- Lassa fever, Bolivian and Argentine hemorrhagic fever viruses (Arenaviridae)
- Congo-Crimean hemorrhagic fever virus (Nairovirus)
- Ebola and Marburg viruses (Filoviridae)
- Hantavirus (Bunyaviridae)
- Rift Valley fever (Phlebovirus)
- Yellow fever virus and Dengue hemorrhagic fever virus (Flaviviridae)

The above group of hemorrhagic fever viruses has not been weaponized to any great extent, though certain agents, particularly Marburg, Ebola (filoviruses), and Lassa fever have been investigated as possible BW agents. Laboratory monkeys have been known to transmit Ebola via the inhaled route, although aerosol transmission among humans has not been confirmed. The high virulence of the filoviruses (for example, Marburg) makes these plausible BW agents.

The virus that causes Lassa fever (named after a small village in Nigeria) was identified in 1969 and killed two-thirds of those who contracted it. Physicians and nurses who treated Lassa fever patients also became ill, even when no direct contact was made with fluids or tissue. Thus it was surmised early on that this virus was easily transmitted, possibly by infectious aerosol, making its suitability as a weapon quite clear.

Except perhaps for Dengue fever, all of the above viruses are known to be transmissible via aerosols or objects contaminated with the virus. Ebola and Marburg (the latter named after the German city where it was first recognized) have been especially lethal in Africa, and a localized outbreak in Germany and Yugoslavia claimed the lives of seven laboratory technicians. While symptoms are shared among the hemorrhagic viruses, clinically they differ in many other respects, and the exact mechanisms by which viruses such as Marburg cause disease are still poorly understood. Symptoms generally are fever, pain, and, in severe cases, hemorrhage from many parts of the body. Death is not caused necessarily by excessive bleeding per se, but the bleeding does indicate the severity of viral penetration.

The former Soviet Union developed the Lassa fever pathogen as a biological weapon and initiated research and development for Ebola and Marburg as biological weapons in 1986. Manufacturing Ebola as a suitable biological weapon was a vexing problem, however, and Soviet scientists had more success with the Marburg virus. The Marburg virus slated for weaponization in the former Soviet Union, Variant U, was originally isolated from a victim of laboratory-acquired infection, Nikolai Ustinov, and subsequently named after him.

Venezuelan Equine Encephalitis (VEE)

The equine encephalomyelitis viruses remain as highly credible threats today, and intentional release as a small-particle aerosol, from a single airplane, could be expected to infect a high percentage of individuals within an area of at least 10,000 km^2. As a further complication, these viruses are readily amenable to genetic manipulation by modern recombinant DNA technology. This capability is being used to develop safer and more effective vaccines, but, in theory, could also be used to increase the weaponization potential of these viruses.[34]

Viruses that can cause inflammation of the brain and associated membranes, particularly the class of encephalitides, could be employed as effective biological weapons. One of these, the Venezuelan equine encephalitis (VEE) virus, often transmitted in nature by mosquitoes, is one of many viral agents under development as weapons at one time or another. Before ending its official BW program in 1969, the United States weaponized the VEE virus, and the Soviet

Union was known to have VEE in its arsenal as an incapacitating-type BW agent. VEE's effectiveness as an aerosol was demonstrated in the former Soviet Union in 1959, when at least 20 laboratory staff became ill with the disease after a vial of freeze-dried VEE virus was dropped and shattered in a stairwell.

In nature, VEE is transmitted by mosquitoes first among horses and other pack animals, creating an epizootic reservoir that can infect human populations. As a biological weapon, VEE can be used in either liquid or dried form and stabilized for effective dissemination.

Symptoms are exhibited in nearly 100 percent of those infected, including inflammation of brain meninges. After a 1- to 5-day incubation period, symptoms include severe headache and fever, along with nausea, vomiting, and diarrhea in some cases. Following a period of rapid-onset fever that can last 1 to 3 days, fatigue continues for up to 2 weeks. Victims of naturally caused VEE infection, especially infants, can develop a serious infection of the central nervous system resulting in convulsions and/or paralysis. While it is debilitating, the mortality rate from untreated VEE is less than 1 percent, although it may be higher for infants and elderly. The number of different strains of the disease, including some with much higher mortality rates, make production of effective vaccines extremely difficult. Data derived from animal studies with VEE aerosols suggest the possibility of higher morbidity and mortality rates, since the inhaled virus can infect the central nervous system in more direct fashion.

Foot-and-Mouth Disease

An outbreak of foot-and-mouth disease in 1997 in Taiwan, a virus that only affects animals (especially cows and pigs), cost the pig farming industry on the island nation some $15 to $25 billion.[35] Though there have been suggestions that Chinese agents introduced FMD in an attempt to destabilize the Taiwanese economy, the available evidence indicates that the source of the infection was the smuggling of meat products and live animals.[36] The rapid spread of the virus throughout Taiwan was a warning, however, that future conflicts or terrorism could utilize the FMD virus to attack a nation's agricultural economy. A small RNA virus (picornavirus), FMD viral particles can be excreted from infected animals and transmitted over large distances. Most critical, however, is the fact that FMD virus can remain virulent for long periods of time and is extremely infectious. The United States, as well as other countries, has studied the possible use of foot-and-mouth disease (FMD) as a BW agent since World War II.

Biological Toxins

The Chemical Weapons Convention (CWC) lists two toxins found in marine organisms (saxitoxin) and in the Castor bean plant (ricin), along with the other Schedule 1 chemical warfare agents and precursors. These two toxins were included as "placeholders" to indicate that biological weapon toxins are also covered by the CWC—at least until a compliance protocol for the 1972 Biological and Toxin Weapons Convention (BTWC) was put into place. Toxins therefore constitute a "gray area" covered by both treaties. The pharmaceutical industry is concerned that the production ceilings on ricin specified in the CWC could limit its use in cancer therapy. Indeed, botulinum toxin was not included in the CWC verification regime because the toxin already has so many legitimate medical applications (for example, for cosmetic treatments and therapy for some neurological disorders).

The category of biological toxins not only includes the mycotoxins listed below, but also those poisons that are produced naturally by living organisms such as microbes, snakes, insects, spiders, and plants. Toxins most likely to be weaponized include botulinum and staphylococcal enterotoxin type B (SEB). SEB is well known as the cause of certain food poisonings, producing painful episodes of vomiting and diarrhea that may last for several hours. These symptoms are caused by the bacterium *Staphylococcus aureus* that produces SEB toxin in the gut, 1 microgram (a millionth of a gram) being sufficient to cause symptoms in humans. Exposure to SEB as an aerosol, however, may have dramatically different and even more deleterious consequences.

Under certain conditions botulinum toxin is produced by the Gram-positive bacterium *Clostridium botulinum* and can cause death in humans by paralysis in extremely small concentration. But this very powerful toxin also finds use as a therapeutic agent in a number of medical applications including the treatment of neurological disorders such as blepharospasm. This is but one example of the dual-use dilemma of biotechnology and commercial applications for potential BW agents.

Although it is conceivable that some biological toxins could be artificially synthesized for BW, it is more likely that toxins would be "harvested" from microbes or extracted from plants. Ricin is an example of the latter: castor oil is produced from the same bean that also contains a small amount of ricin in the remaining mash, about 3 to 5 percent by weight.

Mycotoxins

Toxins produced by fungi, i.e., mycotoxins, are considered possible BW agent candidates and have been controversial for decades, including the putative agent in "Yellow Rain." Trichothecene (T2) toxin is among the most poisonous of the known mycotoxins. Because the *Fusarium* toxin's effects in animals and man are quite similar to those by ionizing radiation (fever, nausea, vomiting), this and other related compounds have been termed "radiomimetic."[37]

Unlike other biological agents, T2 is quite toxic when applied to unprotected skin. But like other biologically derived toxins, related substances have been investigated for possible medical uses, including the treatment of some cancerous tumors. A mycotoxin from *Fusarium equiseti* (diacetoxyscirpenol, or DAS) was considered for an anti-tumor drug formulation called Anguidine, but it was found to be far too toxic for therapeutic use.

The toxicity of another relatively common mycotoxin was clearly demonstrated in the United Kingdom four decades ago, when over the course of a three-month period in 1960 approximately 100,000 turkeys were poisoned. Aflatoxin, considered one of the most carcinogenic substances known to man, was found to be the causative agent in this episode of "Turkey X disease."[38] The discovery of Iraqi production of aflatoxin points to the possible use of a carcinogenic substance in war. According to UNSCOM personnel, Iraq intended to use aflatoxin—mixed with riot control agents—against rebellious factions within the Kurdish minority.[39] Iraqi development of an aflatoxin weapon may have been due to bureaucratic inertia; Iraqi weapons scientists merely wanted to show Saddam Hussein that they had accomplished some concrete results.[40] In the event, as a BW agent, aflatoxin does not offer much in the way of immediate effect on humans, and the Iraqi use of aflatoxin as a possible biological weapon will probably remain unique.

Fungi (Molds)

Before the availability of effective treatment, *Coccidioides immitis* was known to cause high mortality (50 percent) in humans and is an example of a fungus that forms hardy spores suitable for aerosols. Knowing that some ethnic groups were more susceptible to infection from certain fungi such as *Coccidioides immitis*, the United States conducted extremely controversial tests in the 1950s, experimenting on African-American subjects with a fungal simulant, *Aspergillus flavus*.

While most fungi do not generally cause fatal disease in healthy humans, they can be used to destroy crops. In the sixth century B.C., the Assyrians put rye ergot fungus in the water wells of their enemies. While the actual effects of using ergot (wheat rust) in such a manner are not clear, the intent was probably to inflict damage upon their enemies' agriculture.

Botulinum Toxin

Botulism, a disease often acquired from the ingestion of spoiled meat in historical times, is derived from the Latin word "botulus," meaning sausage. As recently as 1973, a large outbreak of botulism from poorly processed vichysoisse forced the United States to finally install federal regulations for canned food processing.

Botulinum toxin is the causative agent in botulism, produced by the bacteria *Clostridium botulinum*, and is released upon the death and breakdown of the organism. In humans, the toxin produced by *Clostridium botulinum* is among the most poisonous substances known.[41] As little as 0.000005 micrograms of botulinum toxin is enough to kill a small laboratory mouse, while 0.07 to 0.10 micrograms may be enough to kill most humans. (0.1 micrograms is 10 millionths of a gram). The toxin (type A) of *Clostridium botulinum* is 100,000 times more powerful than the sarin nerve agent. However, the deadly botulinum toxin works in a reverse manner. Instead of increasing the level of accumulated neurotransmitter (i.e., acetylcholine), botulinum toxin *blocks* the release of acetylcholine and thereby prevents transmission of nervous impulses. Botulinum's activity on muscle and nerves has found medical applications, particularly for the treatment of dystonia, a category of neurological disorders.

Classically, botulism in humans occurs as food poisoning, including infant botulism from the ingestion of spore-laden honey.[42] Other types of botulism occur in nature in ducks and geese following the anaerobic fermentation of plants by the bacteria *Clostridium botulinum*. Poisoning in humans does not involve a fever, although it does produce difficulty in breathing and problems with vision, and the pupils may become fixed. Death is caused by respiratory arrest that can follow within 24 hours after ingesting the toxin, although the mortality rate from botulism can be reduced to 10 percent with adequate treatment.

Following a biological attack in the form of an aerosol, signs of botulinum poisoning might appear much faster than in the food-borne form of illness. While the battlefield scenario includes inhaled particles, the use of projectiles such as flechettes contaminated with botulinum toxin is another possible delivery method.

Staphylococcal Enterotoxin Type B (SEB)

Most people encounter the bacterium *Staphylococcus aureus* and its toxin staphylococcal enterotoxin type B (SEB) at some point in their lives from food (ptomaine) poisoning. Some of these strains of bacteria have been responsible for toxic shock syndrome among women using feminine hygienic products, especially during the peak of the affliction in the early 1980s.

As a potential biological weapon, aerosolized SEB toxin would present a significant hazard. Inhaling SEB could cause 80 percent or more of targeted personnel to become extremely ill within 3 to 12 hours, with possibly even more dire consequences than other routes of intoxication. The toxin is relatively heat-resistant, and very small quantities can produce severe illness. These properties give rise to concern in terms of a possible attack using SEB in water or food supplies. It might take up to two weeks for humans to recover from SEB poisoning, and higher concentrations of SEB could cause septic shock and death.

Ricin and Saxitoxin

Both ricin and saxitoxin are listed under the Chemical Weapons Convention Schedule 1 as chemical weapons, although both can be found in nature. (Saxitoxin can be manufactured synthetically in small quantities).

Ricin is a protein toxin extracted from the castor plant (*Ricinus communis*) and, almost like a virus, can insinuate itself into cells, disrupting as it does the process of cellular protein manufacture. Ricin was used by the Soviet KGB in 1978 to assassinate the Bulgarian dissident Georgi Markov in London. Iraqi military scientists experienced great difficulty in weaponizing ricin, indicating its use on the battlefield may be less a threat than SEB or botulinum toxins. (The latter agents are considerably more toxic than ricin in any case). Nonetheless, as a possible BW agent, ricin is considered a real threat, and research continues in the United States to develop prophylactic treatment and vaccine for it.

Saxitoxin is related in structure and effect to that of the Japanese puffer fish. Being a neurotoxin, saxitoxin is derived from so-called dinoflagellates that are usually found in red tides.[43] When shellfish used in seafood feed on these organisms, they can become dangerous carriers of toxin. After consumption of the toxin, paralytic shellfish poisoning can result; untreated humans will die within 2 to 12 hours. (Reportedly, the U-2 spy aircraft pilot Francis Gary Powers carried a hollowed-out silver dollar containing saxitoxin as a method of suicide.) Battlefield weapons using saxitoxin might employ small flechettes fired from rifles, or specially-designed artillery shells such as the 105-mm bee-hive round.

In the United States saxitoxin was stockpiled until 1970, when President Nixon declared that BW toxins would be eliminated from the US offensive arsenal. However, 10 grams of saxitoxin remained in Washington, DC, as late as 1975. Because of its potential for medical research, this quantity of saxitoxin was finally distributed to health researchers by the National Institutes of Health.

Trichothecene Mycotoxins (T2)

Decades ago in the former Soviet Union, improperly grown and cultivated wheat led to mold (*Fusarium*) infestation of the grain. Due to the toxin produced by the fungus trichothecene (and others), thousands were stricken with alimentary toxic aleukia (ATA), a condition reported by Russian scientists in

1943. Typical ATA symptoms include vomiting, severe skin irritation, and internal bleeding. During the years 1942–47, over 10 percent of the population in a major district near Siberia died from moldy grain, and trichothecene toxin (T2) was subsequently implicated. Allegations that the Soviet Union and her allies were using a BW toxin, T2 or trichothecene mycotoxin, in southwest and southeast Asia during the 1970s and 1980s have not yet been confirmed. Similar conditions that gave rise to ATA from trichothecene have occurred elsewhere, such as in Japan (Red-Mold disease) and others in North America. More recently, *Fusarium* is reported to be a serious fungal infection for patients whose immune systems are severely weakened, such as those undergoing aggressive cancer treatment.

Trichothecene mycotoxin (i.e., toxin derived from fungi) is unique among the BW agents in that it is immediately active upon contact with the skin. There are a number of fungal genera that produce T2 or similar toxins, the *Fusarium* mold species being the best-known source. The T2 toxin stops protein synthesis within cells and has been found to be highly toxic to tissues, especially the skin. T2 targets with particular affinity those cells that are replaced rapidly, such as the skin, mucous membranes, and bone marrow.

T2 would be a particularly effective against unprotected civilians or military personnel by causing enormous pain, general discomfort, and, in large enough concentrations, agonizing death. Trichothecene mycotoxin would also present a contamination hazard for clothing and equipment.

BIOREGULATORS

Bioregulators are compounds that can radically alter the body's physiological state. They may also be pursued and developed by some BW programs. Analogous to the Lipid A toxin found in bacteria, "Substance P," an 11-amino acid molecule (peptide) has been demonstrated to lower blood pressure dramatically in quantities as small as 1 microgram or less. These substances, however, are somewhat exotic in the context of bioweaponry, and most proliferators are unlikely to go much beyond the stage of research with these agents.

PROTOZOA

Forming a category unto themselves, the protozoa are perhaps the least likely of BW agent candidates, even though they cause such diseases as giardiasis, toxoplasmosis (*Toxoplasma gondii*), and schistosomiasis (blood flukes). Except for individuals with severely weakened immune systems, common protozoa such as those that cause giardia or cryptosporidium usually do not pose a death threat to humans, although they certainly can make many people physically ill to the point of being incapacitated. By and large, most BW experts do not give protozoa serious consideration as possible biological weapons. However, recent cases, such as the cryptosporidium outbreak in Milwaukee in 1993 that caused 111 deaths and some 400,000 casualties, may provide the occasion for a second look.

CHAPTER 8

Biological Warfare: A Brief History

We have defined biological warfare as the use of pathogens, that is, disease-causing bacterial and viral agents, or biologically derived toxins against humans, animals, or crops. The very word "toxin, " from the Greek for "bow," originated from the use of poisons on arrowheads,[1] showing that humans have utilized disease and toxins in combat throughout history.

BIOLOGICAL WEAPONS IN ANCIENT TIMES

As long ago as 400 B.C., Scythian archers dipped their arrows in feces and putrefying corpses, predating by thousands of years the Viet Cong use of *punji* stakes coated with human excrement as booby traps. Similarly, Roman soldiers ran their swords into manure and the rotting offal of dead animals before battle. If wounded by a weapon contaminated in such fashion, victims contracted infection (especially tetanus) as a result. History also records the employment of poisonous snakes as a kind of toxic delivery system—tossed onto King Eumenes's ships by Hannibal's men during the Second Macedonian War (190 B.C.).

The siege of the Black Sea city of Kaffa (now Feodosia, Ukraine) by the Mongols in A.D. 1346–7 is frequently cited as an example of BW. But a close examination fails to show to what extent disease employed as a weapon contributed to the final outcome. A heavily defended, walled city populated by merchants and traders from the Italian city of Genoa, Kaffa was surrounded by a Mongol horde but remained impregnable. This reflected the fact that at least until the development of gunpowder and artillery it was more difficult to defeat a fortified castle than to defend it. More often than not in the Middle Ages investing or surrounding a fortress was followed by months, even years of stalemate. In addition, the attacking forces camped outside the castle had to deal with the various pests and vermin they had brought with them and the diseases that broke out as a result. In the case of Kaffa,

> ... [T]he Tartars died as soon as the signs of disease appeared on
> their bodies: swellings in the armpit or groin caused by coagulat-
> ing humors, followed by a putrid fever. The dying Tartars, stunned
> and stupefied by the immensity of the disaster brought about by
> the disease, and realizing that they had no hope of escape, lost
> interest in the siege. But they ordered corpses to be placed in cat-
> apults and lobbed into the city in the hope that the intolerable
> stench would kill everyone inside. . . .[2]

According to legend, it was out of frustration, with Bubonic plague rampant among his own men, the Mongol khan Janibeg ordered that the plague-
infected corpses of own troops be catapulted over the walls into Kaffa. Some
historians believe that a resulting plague epidemic weakened Kaffa's capacity
for self-defense and led to its eventual downfall.[3] According to others, it was a
retreat by plague-ravaged Tatars themselves that ended the siege in 1347. In
any event, when the Genoese returned to their native city via Venice, they
were followed by flea- and plague-infested rats. The subsequent peregrinations
of the Genoese no doubt contributed to the spread of Black Death in Europe,
which would ultimately lose at least a fourth of its population to the plague.

It should be noted that the technique of tossing plague victims' corpses
into the midst of an enemy was probably not an effective way to spread the
disease. While *Yersinia pestis*, the causative organism of plague, can remain
infectious in body fluids after a victim's death, the bacteria are dependent
upon a vector, the flea, to spread. As a body cools, fleas quickly jump off to
seek another warm living host for blood. At the time of the Kaffa siege, the
population of rats around the Black Sea was quite large and could have served
as a reservoir of plague bacteria.

Biological Warfare in the New World

Although there is no evidence that Europeans employed BW against the
indigenous populations of America before the 1700s, the natural course of disease transmission had already done much of the work. Because the natives of
the Western Hemisphere had no resistance to Old World illnesses, the ravages
of smallpox and other infectious disease carried by the Europeans helped pave
the way for colonial expansion in the Americas. It is estimated that, beginning
with the first contacts with Europeans at the opening of the sixteenth century,
as many as 95 percent of Native Americans were ultimately wiped out by
imported germs.

Though this devastation appears to have been inadvertent, there is strong
evidence of a campaign to use smallpox against Native American Indians during the French and Indian War (1754–1760).[4] Two British officers, Colonel

Henry Bouquet and General Jeffrey Amherst, undertook to spread smallpox among their Indian foes in the hope of achieving "the Total Extirpation of those Indian Nations."[5] British traders were enlisted in a scheme to give the Indians blankets and clothing taken from a hospital that treated smallpox victims. Though this plot appears to have had only limited effect, it is a milestone in the history of biological warfare.

In 1775, at the beginning of the American Revolution, the talented, later infamous military commander Benedict Arnold led his men from Fort Ticonderoga in an invasion of Canada. But Arnold's valiant effort to capture Quebec ended in failure, and he was forced to retreat. His troops were convinced that the British had spread smallpox among them, perhaps through civilians who had socialized with Continental troops. (This claim was echoed by Thomas Jefferson who accused the British commanding officer in Quebec of "sending" smallpox into the American army.[6]) In April 1775, General George Washington, who had spent considerable time and effort emphasizing the importance of hygiene in his own army, received intelligence that the British were attempting to use smallpox against the rebellion:

> . . . [T]he information that I received that the enemy intended
> Spreading the Small pox amongst us, I could not Suppose them
> Capable of—I now must give Some Credit to it, as it has made its
> appearance on Severall of those who Last Came out of
> Boston. . . . [sic][7]

While smallpox was blamed, among other things, for the failed expedition against Quebec, there is no proof that the British attempted to use it. Indeed, indications are that smallpox was a self-inflicted malady for the Continental soldiers. At one point, many soldiers in the Continental Army took it upon themselves to inoculate one another with smallpox, hoping thereby to avoid the disease. This ad hoc procedure probably helped spread the contagion. This episode may have convinced Washington of the necessity of effectively inoculating his troops against the disease in 1777.

Later in the Revolutionary War a plan was suggested to the British commander General Charles Cornwallis for "distributing" some 700 blacks infected with smallpox among the rebel-held plantations. The results of this particular plan, if in fact it was ever carried out, are unknown. It would seem that in this period at least, utilizing disease as a weapon was more along the lines of psychological than veritable biological warfare.

BIOLOGICAL WARFARE IN MODERN TIMES

While there are scattered instances of the deliberate use of disease-causing agents in warfare before the nineteenth century, a systematic approach to BW had to await a true understanding of the nature of disease itself. Only in the nineteenth century was the germ theory of disease advanced and accepted. Techniques to isolate, grow, and purify microbial cultures were adequately developed in the same period, following upon Louis Pasteur's observations, Joseph Lister's work with understanding the cause of infection in 1878, and Koch's postulates codifying the definition of disease-causing microbes. It was not until 1900 that bacteria were finally being classified.

Therefore it is difficult to state with certainty that certain outbreaks of disease in history were the result of human decision-making, and distinguishing between natural epidemics and those brought about by biological weapons remains difficult. Nevertheless, by World War I, the biological sciences were advanced enough for the identification of certain pathogens to lead to their being isolated and cultured.

World War I

In wartime Germany, scientists and military strategists often discussed the use of anti-personnel BW agents, but its use against humans was repeatedly and resoundingly rejected by the German government. When an enthusiastic military hygienist recommended dropping plague on Great Britain from dirigibles, the official reply from his superiors was unequivocal: "All respects to your courage and patriotism, but if we undertake this step we will no longer be worthy to exist as a nation."[8]

This restriction on BW agents against humans did not extend to anti-animal or anti-crop agents, however. In World War I, German secret agents employed at least two bacterial pathogens, the causative agents of glanders (*Burkholderia mallei*) and anthrax (*Bacillus anthracis*), infecting horses and cattle destined for use by Allied forces at the front.

In 1915, a German-American physician, Dr. Anton Dilger, produced BW agents in his Washington, DC, home, culturing anthrax and glanders from seed stock provided to him by the Imperial German government. Baltimore stevedores were paid by German agents to infect some 3000 mules, horses and cattle bound for the Allies in Europe.

While one source reports that "several hundred military personnel" were affected by this operation,[9] conclusive evidence has not been found. Although glanders can cause chronic disease in humans, no reports are available of any infections among Allied personnel. It also remains a curiosity that, despite the very infectious nature of the pathogen, none of the American agents recruited by Germany to infect the animals became sick themselves.

Evidence suggests that Germany itself was the target of BW sabotage during World War I. In one particular instance, German intelligence became aware of French operatives in Switzerland infecting horses bound for Germany with *Burkholderia mallei*.

The efforts at BW sabotage on both sides may have been so inconsequential or knowledge of these activities too compartmentalized for subsequent governments to have been fully aware of them. Whatever the reason, it is noteworthy that among the many military restrictions imposed upon Germany by the Versailles Treaty (1919), including the proscription of "asphyxiating, poisonous or other gases," biological weapons were never mentioned.[10]

The Geneva Protocol of 1925

The abhorrent image chemical weapons gained during World War I gave momentum to the establishment of a treaty "for the Prohibition of the Use in War of Asphyxiating Gases and of Bacteriological Methods of Warfare."[11] Due to lobbying by the Polish delegation during the 1925 Geneva Conference, the proscription of BW was added. Anticipating the BW proliferation concerns that would emerge 75 years later, the originator of the initiative, a Polish general, Casmir Sosnkowski, warned: "The bacteriological weapon can be manufactured more easily, more cheaply and with absolute secrecy."[12]

In this, the first diplomatic effort to limit the use of biological weapons, the 1925 Geneva Protocol prohibited the deployment, but not the research, production, or stockpiling, of BW agents; in addition, some countries reserved the right to retaliate with BW if they were attacked. Despite the addition to the Protocol concerning BW, in the 1920s biological weaponry was not a widely perceived threat. Referring to a League of Nations committee formed on the subject, the US Chemical Warfare Service wrote in its 1926 annual report that BW did not seem to be a viable method of waging war. This is worth quoting at length, for it touches upon just about every basic element of modern biological weaponry. It concluded that biological weapons

> ... would have little effect on the actual issue of a contest in view
> of the protective methods which are available for circumscribing
> its effects. The pollution of drinking water by cultures of typhus
> or cholera germs would be combated by filtering, as already prac-
> ticed in large centers, or by treating the waters of rivers with
> chlorine. The enemy would have to contaminate, by means of air-
> craft, the filtered water of the reservoirs directly; this would be a
> difficult operation and its effects could be frustrated by preventive

vaccination. The propagation of plague by pest-infested rats would be as dangerous for the nation employing this method as for its adversary, as rats pass freely between the lines of both armies. Experience has shown, moreover, that it is possible speedily to check an outbreak of plague. Moreover, the danger of an epidemic of typhus propagated by lice has greatly diminished. As regards the poisoning of weapons, the experts point out that the germs which could be employed (streptococci, staphylococci, anthrax spores, glanders bacilli, etc.) would not preserve their dangerous properties if they were prepared a long time beforehand and allowed to dry on metallic surfaces. Nor if placed in a projectile would these germs better resist the shock of discharge, the rise of temperature and the violence of an explosion which destroys all life. The only method presenting a certain danger would be that of dropping from aeroplanes, glass globes filled with germs. Finally, the majority of the experts are of opinion that bacteriology cannot at present produce infective substances capable of destroying a country's live stock and crops. . . .[13]

Up into the 1930s, the general consensus in Germany was that biological weapons were impractical, and furthermore BW could present a "boomerang" threat to attacking forces. Neither had informed opinion in America substantively changed since the original Chemical Warfare Service findings ten years earlier. In 1933, for example, Major Leon Fox of the US Army maintained that more modern sanitary measures would effectively counter a biological weapon threat against the United States.[14]

Japanese BW, 1932–1945

One of the few countries known to have engaged extensively in modern biological warfare, Japan conducted an ambitious biological weapons program in occupied Manchuria between 1937 and 1945. Known as Unit 731, the Japanese organization consisted of a laboratory complex of 150 buildings and five satellite camps, employing over 3000 scientists and technicians in Harbin, China.

The Japanese experimented on prisoners with plague, cholera, epidemic hemorrhagic fever , the effects of frostbite, and even some sexually transmitted diseases. During the Japanese program, at least 10,000 prisoners died as a result of infection or execution following experimentation. It is difficult to calculate how many Chinese civilians in the surrounding areas died as a result of field testing and outright biological attacks by Japanese Army scientists. Some sources in China have claimed that these could number in the hundreds of thousands.[15]

In 1940–1, the Japanese dropped ceramic bombs containing fleas infected with bubonic plague over Manchuria and other regions of China. The bombs carried grain to attract rats which were bitten by the infected fleas, thereby transmitting plague bacteria (*Yersinia pestis*) to the human population. Japan attacked more than eleven Chinese cities with various biological agents via food, water, or agricultural sprayers. In Chang De, for example, a municipality with no previously recorded cases of plague, two weeks after a Japanese plane was spotted making low passes over the city, six people came down with the disease. The Japanese also stockpiled 400 kilograms of anthrax for use in specially designed fragmentation bombs. Still, most Japanese biological attacks were generally not successful; in at least one instance a release of cholera bacteria into a river resulted in the infection of Japanese troops, causing 10,000 casualties and 1700 deaths.

While the Japanese themselves acknowledged that some 20,000 Chinese died as a result of BW operations, not surprisingly, Chinese estimates of deaths caused by Japanese BW activities are much higher. According to one PRC (People's Republic of China) publication on arms control, "During Japan's invasion of China, BW was carried out among 20 or more provinces and cities in China, causing more than 200,000 casualties among the Chinese people."[16] Japanese and Chinese scholars have since come to the conclusion that "at least 270,000 Chinese soldiers and civilians were killed as a result of Japanese germ warfare between 1933 and 1945."[17] However, no information to date can support such a figure, nor is it likely that Japanese BW activities could be definitively linked to every occurrence of plague or other illnesses. (Plague, for example, has been endemic to China since 1894, and during wartime it is not uncommon for authorities to lose control of infectious diseases.)

United States Congressional hearings held in 1986 were unable to ascertain if Allied prisoners were among the many victims of Unit 731. However, a record exists of at least one attempt to attack American forces in the Pacific. To counter US island-hopping campaigns underway in 1944, General Ishii Shiro, the director of the Japanese BW program, sent a ship off to Saipan with about 20 men armed with biological weapons, including plague, but along the way, the ship was sunk by the US Navy. The Japanese also considered the use of plague to resist the American invasion of Okinawa in 1945, although details on how it was to be carried out are stretchy at best.

Following Japan's surrender in 1945, General Ishii ordered Unit 731 burned to the ground, but the United States granted amnesty to Japanese scientists involved in BW research on the condition that they disclose all of their data, including the results of human experimentation. The United States and Soviet governments held hearings (some secret) in which details of Ishii's work were further disclosed.

THE UNITED STATES BW PROGRAM

Shortly before the beginning of World War II, in part because of a growing sense of potential conflicts brewing in Europe, and to a lesser extent Asia, the US government's interest in biological agents began to revive.

Development Phase: 1939–1950

In the period just preceding World War II, although the United States remained confident that modern sanitary methods could counter most biological weapon threats, intelligence reports indicated that Japan and Germany had undertaken research into offensive BW preparations. In February 1939, the US State Department reported that a Japanese Army physician in New York tried to obtain yellow fever virus, possibly an attenuated strain, from the Rockefeller Institute for Medical Research.[18] This and other intelligence contributed to a new sense of urgency, and in September 1939, American military scientists decided to reexamine the problem of BW. A Chemical Warfare Service report that followed listed nine diseases and pathogens that could be potential BW agents: "yellow fever, the dysenteries, cholera, typhus, bubonic plague, smallpox, influenza, sleeping sickness [*Trypanosoma brucei*, via the Tsetse fly], and tetanus." These agents were of great concern because they could be spread by insects, or otherwise did not require "existing skin lesions nor a co-agent in order to enter the human body."[19] Although doubts still persisted in America as to the real threat of BW, a consensus was reached for the need of funds for research into defenses against biological weapons.

In February 1942, the National Academy of Sciences submitted a report to Secretary of War Henry L. Stimson, describing the threat posed by BW agents against crops, livestock, and humans and concluding with a call for both defensive and offensive BW research. Stimson, in turn, recommended to President Franklin Roosevelt that the report be adopted. According to Stimson's diary, at a May 15, 1942, cabinet meeting, Roosevelt said that he had not read the report but ordered its implementation nevertheless. In mid-1942, George W. Merck, President of Merck & Co., Inc., was made chairman of the War Research Service (WRS) established to oversee US BW-related activities, while the CWS was given the responsibility of building and operating laboratories and production facilities.

In March 1942, the CWS suggested that, in addition to work conducted by civilian research scientists in biological weapon defense, the following BW agents be studied in an offensive context:

- *Anti-personnel agents:* anthrax, psittacosis, plague, cholera, typhus, yellow fever, coccidioidomycosis, typhoid, and paratyphoid.

The first biological warfare laboratory built at Camp Detrick (later Fort Detrick), Maryland. (Courtesy of Soldier Biological and Chemical Command, Historical Research and Response Team, Aberdeen Proving Ground, MD.)

- *Anti-animal agents:* rinderpest, foot-and-mouth disease, and fowl plague (Avian influenza).
- *Anti-crop agents:* rice blast, wheat rusts, and South American rubber leaf blight.

By 1944 this list would also include:[20]

- *Anti-personnel agents:* tularemia, brucellosis, glanders, and Melioidosis.
- *Anti-crop agents:* late blight of potatoes and sclerotium rot.

Meanwhile, in 1942–3, the CWS constructed a BW facility at Camp Detrick (later renamed Fort Detrick), an army facility in Maryland, at a cost of $13 million. Operational in 1943, the facility at Camp Detrick employed approximately 4000 people. Other BW-related facilities included a 250-acre site near Dugway Proving Grounds (Utah) and a 2000 acre facility at Horn Island (Pascalouga, Mississippi), both of which were used for open-air testing.

At the same time, cooperation in BW research and development was undertaken by the United States, Canada, and Great Britain. It was learned from tests conducted by the British at Gruinard Island off the coast of Scotland and Penclawdd on the coast of Wales (1942–3) that loading anthrax into bomblets included in cluster munitions was the most feasible method of delivering the agent. This was determined by sheep placed at various distances from a bomb loaded with anthrax. The reach of the deadly spores was such that animals placed 250 yards downwind received a lethal dose of anthrax. Despite these impressive results, effective cluster-type munitions using anthrax were never supplied to Allied forces in World War II.

In the United States, BW agent simulants were manufactured at a 6100-acre site at the Vigo Plant in Terra Haute, Indiana. Originally built to produce conventional ordnance, the Vigo Plant was converted into a testing facility for BW agents, but safety concerns limited large-scale production to innocuous microbes only like *Bacillus globigii* (BG); even then, full-scale production did not begin until June 1945. BG and another bacteria, *Serratia marcescens* (SM), are relatively innocuous bacteria, used to simulate how BW agents might spread in different environments, and for testing structural barriers such as gas mask filters. *Serratia marcescens* was especially effective as a simulant because its distinctive red color made it easier to survey. By war's end, some 8000 pounds of BG had been produced before the project was shut down. (In 1947, Pfizer Inc., the pharmaceutical giant, purchased the Vigo installation at a US government auction and still utilizes the 20,000-gallon fermenters to produce veterinary antibiotics.)

By April 1943, Britain had manufactured some 5 million linseed cakes at a soap bar factory in London and injected each of these with about a half a milliliter of anthrax slurry. These were designed for retaliation in the event of the use of BW by Germany, the plan being to distribute them over enemy cattle-grazing grounds. The idea was to wipe out Germany's supply of beef, not necessarily to spread anthrax to civilians who would eat the infected meat from infected cattle (although this too was a theoretical possibility).

The Post-World War II Era and the Korean War

These bandits in generals' uniforms, the butchers in white gloves, the bloody bigots and traders in death who have unleashed the most inhuman carnage in history, warfare with the assistance of microbes, fleas, lice and spiders.

—Pravda, *1952*[21]

In the period following World War II, US BW production capacity was gradually reduced to laboratory scale research and development. In 1947–9, small-scale, open-air testing of simulants, *Bacillus globigii* and *Serratia marcescens*, was carried out at Camp Detrick. Pathogen tests began at Camp Detrick in 1949 in an enclosed, 1-million-liter steel sphere called the "eight ball."

During the Korean War, the United States expanded its BW program. The government established the Pine Bluff Arsenal BW agent production facility in Arkansas, and enlarged the research facilities at Camp Detrick. By 1951, the program successfully developed, tested, and produced a variety of anti-crop agents for military purposes, as well as bombs capable of delivering such agents. During fiscal years 1951–3, spending on BW-related research and development amounted to more than \$345 million (while funds allocated for chemical weapons were approximately \$420 million in the same period). However, by September 1952, the ability to deliver biological agents effectively was still an open question, and it had become apparent, at least to the United States Air Force, that it still had no "highly lethal, stable, viable, easily disseminated, low cost, epidemic-producing, BW agent."[22]

In March 1951, during the Korean War, Peking radio charged that the United Nations Command—the US, the South Koreans, and their allies in the war—was manufacturing biological weapons for use against North Korea. In May of that same year, North Korean Minister of Foreign Affairs, Pak Hen Yen, protested to the United Nations that the United States had attacked Pyongyang, the North Korean capital, with smallpox. Almost a year later, on February 22, 1952, North Korea made more detailed charges, that

...The American imperialist invaders, since January 23 this year [1952], have been systematically scattering large quantities of bacteria-carrying insects by aircraft in order to disseminate infectious diseases over our front line positions and rear. Bacteriological tests show that these insects scattered by the aggressors on the posi-

tions of our troops and in our rear are infected with plague,
cholera and the germs of other infectious diseases. This is
irrefutable proof that the enemy is employing bacteria on a large
scale and in a well-planned manner to slaughter the men of the
[Korean] People's Army, the Chinese People's Volunteers, and
peaceful Korean civilians. . . .[23]

In response to these charges, the UN commander General Matthew B.
Ridgeway addressed the United States Congress, saying "no element of the
United Nations Command has employed either germ or gas warfare in any
form at any time."[24] Meanwhile, the Soviet Union joined the North Koreans
and the Chinese rejecting a proposed investigation by the International
Committee of the Red Cross, accusing the organization of being a "lackey of
American imperialism" obviously out to "whitewash the perpetrators of the
crime with a worthless report."[25] This was followed up on March 8, 1952, by
Chinese Premier Zhou Enlai's claim that the United States had bombarded
both Chinese and North Korean territory with disease-laden chicken feathers, shell-fish, rats, and insects. (The contamination of clams was supposedly
caused by the US Air Force's destruction of a water purification plant nearby.)
A team of scientists was sent by the International Scientific Commission, an
arm of the pro-Soviet World Peace Council, to investigate the BW charges in
March 1952. After the visit and tour of the area involved, the team declared
the North Korean charges credible.

A year later, in April 1953, the Soviets insisted that the failure of the United
States to ratify the Geneva Protocol of 1925 was further evidence that the US
did in fact use BW in Korea. The granting of immunity by the US to Japanese
personnel involved with Unit 731 was also used to buttress the Communist
allegations.

Contrary to these accusations, a US Joint Strategic Plans Committee report
of 1953 that addressed America's chemical and biological warfare preparedness
stated the United States was far from being capable in the area of biological
weapon delivery. As a result, the report contended the United States had only
a limited capability in antipersonnel and anti-crop BW and no capability at all
in anti-animal BW.

The only "evidence" corroborating accusations that the United States used
biological weapons in Korea were statements made by American prisoners of
war after being "brainwashed" by their captors. In this regard, to this day, a
North Korean museum displays handwritten "confessions" from captured US
servicemen who "admitted" that the United States used germ warfare.

Documents coming to light since the end of the cold war have revealed the
sequence of events that led to Chinese/North Korean allegations of US
biowarfare in Korea.[26] In 1952, the North Korean health minister went first to

Beijing, then to Mukden (Manchuria) where he obtained *Yersinia pestis* (plague) bacteria. Two condemned North Korean prisoners were infected with the plague. One of them was put to death and his body used as "evidence" of biological warfare, which was presented to the visiting International Scientific Commission team mentioned above.

Documents coming to light in the post-Cold War era demonstrat a concerted effort by North Korea and China to fabricate evidence pointing to the use of biological weapons by the United States.[27] Among these was a report made for the eyes of Lavrenti Beria, the notorious head of Soviet intelligence at the time of the Korean war, "False plague regions were created," and "burials . . . were organized, measures were taken to receive the plague and cholera bacillus."[28] The campaign ended at about the time of Joseph Stalin's death in 1953, when the Korean war was winding down.

US Testing Activities, 1951–1969

Ironically, given the baselessness of Communist accusations, the US did make some significant advances in its BW capacities during the Korean War. In Operation Dew (1951–52), 250 pounds of a fluorescent tracer were released from a minesweeper off the southeast coast of the United States to study the behavior of aerosols. Another exercise, Brown Derby, was carried out in November 1953 by the Chemical Corps and US Air Force to assess the ability of the US to produce and transport BW weapons overseas. A simulant was manufactured at Pine Bluff Arsenal, inserted into cluster bombs, then shipped to Eglin Air Force Base in Florida. The exercise demonstrated the US ability to mount a BW operation from scratch in a matter of days.

A series of at least three tests, code-named "Bellwether," to study the biting behavior of mosquitoes, was carried out beginning in the late 1950s. During "Bellwether 1," for example, in September–October 1959, uninfected, female *Aedes aegeypti* mosquitoes were released in 52 field trials, and the number of bites on laboratory animals and humans were tallied.

Operation Big Itch was a series of field tests of E–23 and E–14 bombs loaded with *Xenopsylla cheopis* fleas. The fleas were dropped over guinea pigs at the Dugway Proving Grounds in 1954. In one test, technical problems permitted fleas to escape into the plane and bite the bombadier, observer, and pilot. Operation Big Buzz involved breeding and loading approximately one million uninfected *Aedes aegypti* mosquitoes into E–14 munitions and discharging them in the state of Georgia.

As the BW development effort expanded, scientists from Fort Detrick secretly performed animal studies at remote desert sites and on barges near Johnston Atoll in the Pacific Ocean. Vulnerability tests performed in the 1950s and 1960s in New York City and San Francisco covertly employed simulants. In order to monitor the progress of bacteria in a possible BW attack, a US Navy minesweeper projected an aerosol consisting of the simulant SM off the coast of San Francisco in the early 1950s. During the BW tests at that time, SM was considered to be relatively safe and had been also used to simulate the behavior of microbes through different environments (including that of hospitals).

Unfortunately, simulants like SM were not entirely harmless, having caused diseases such as pneumonia and septicemia in some patients. A case in point was the death of Edwin Nevin, a 75-year old patient who appeared to die from an SM infection following a vulnerability test in San Francisco (1950s). Nevin's family sued the US Government for damages in 1980, but the courts finally ruled against the family on the basis that the tests were performed in order to protect national security.[29]

Eleven (including one fatal) infections involving SM occurred at Stanford University Hospital, California, from September 1950 to February 1951. It was not clear, however, that these cases were actually related to a vulnerability test in the area. An expert panel concluded that the aerosol tests were in fact unrelated, and that continued use of SM would not represent a high risk to the public. Between 1949 and 1968, the US government surreptitiously dropped BW simulants (such as *Serratia marcescens*) over a number of American cities, including San Francisco and New York City, for the purpose of testing vulnerability to biological attack, as well as possible weapons and delivery systems. In 1955, American scientists and military experts began using human volunteers to test the effect of various BW simulants including the anthrax simulant *Bacillus globigii*.

Despite increasing public interest in disarmament during the 1960s, the American offensive BW program continued to grow. Steps included the development of large-scale freeze- and spray-drying systems to improve the survivability of biological weapons. Research on various insect vectors was conducted as well. By 1966, government facilities at Pine Bluff Arsenal and Fort Detrick had already mass-produced several BW agents for use in several types of munitions. By the time of termination of its BW activities in 1969, the US had seven standardized biological weapons. In the lethal category were the bacterial agents that cause anthrax and tularemia. Incapacitants included the causative agents of brucellosis, Q-Fever, and VEE. Also weaponized were the lethal botulinum and SEB toxins.

In 1969 and 1970, President Richard Nixon issued National Security directives renouncing all offensive development and production of microbial and toxin agents. By 1972, all anti-personnel BW agent stocks and munitions were destroyed. The United States also terminated all offensive research, closed or cleaned up all offensive facilities, and turned them over to other government agencies for other research. The unilateral disarmament initiated by President Nixon's directives set the stage for the 1972 Biological and Toxins Weapons Convention (BTWC). On January 22, 1975, the United States finally ratified the BTWC on the prohibition of the development, production, and stockpiling of bacteriological and toxin weapons.

SOVIET BIOLOGICAL WEAPONS: 1919–1989

While the formal military Soviet BW program began in 1946, its origins go back much further, to Lenin's establishment of the Bacteriological Institute in 1919 in Saratov. Epidemic typhus following World War I affected about 30 million Russians, and some 3 million people died in the years 1918–1922.[30] The typhus epidemic convinced Soviet leaders that, if properly harnessed and designed, biological weapons could have a devastating effect against their enemies. During the early years of Soviet biological weapons development in the late 1920s and in the 1930s, infected animals were killed, dried, and their carcasses ground into powders for use in biological weapons. A "secret bacteriological institute" was also founded in Suzdal in the 1930s, where research was conducted into using tularemia and plague bacteria as BW agents, as well as *Rickettsia prowazekii* (epidemic typhus), tularemia, and Q-Fever. Techniques to produce the causative organisms of smallpox, plague, and anthrax for biological weapons were also studied in years prior to World War II. Actual weaponization included a few pathogens, reaching to about 10 after the war. But further developments in BW research would have to await decades of turmoil and paralysis in the Soviet scientific community, due to Stalinism and the cult of Stalin's henchman, the pseudoscientist Trofim Lysenko.

The Soviet Renaissance (1973–1989)

Not until two decades after the death of Stalin and the eventual loosening of the grip of Lysenkoism did Soviet science recover. In 1972, Yury Ovchinnikov, vice president of the Soviet Academy of Sciences and a molecular biologist, took the case for biotechnology to the Soviet Ministry of Defense and was given a green light to embark on a renewed and intense effort to develop BW agents. Spurred by Ovchinnikov, new biological weapons were added to the Soviet arsenal, including weaponized anthrax, smallpox virus, and plague bacteria.

In April 1979, the Soviet BW production complex in Sverdlovsk accidentally released anthrax spores that killed at least 66 people downwind (other estimates go as high as the hundreds or even thousand). Soviet officials claimed that the outbreak of anthrax occurred because people ate contaminated meat purchased on the black market. However, autopsies revealed that the victims suffered symptoms unique to inhalation anthrax.

By the early 1990s, US intelligence was able to prove what it had suspected all along: The Soviet Union had been producing weaponized bacteria at Sverdlovsk, in clear violation of the Biological and Toxin Weapons Convention (BTWC) which it had signed and ratified years before in 1972.

In the former Soviet Union, hundreds of tons of weaponized *Bacillus anthracis* were eventually produced and stockpiled. The BW facility at Stepnogorsk in Kazakhstan once used 20,000-liter-capacity fermenters to grow anthrax spores. Similarly, thousands of pounds of smallpox and plague were also produced as strategic biological weapons. Smallpox was produced in liquid form and was intended for delivery in cantaloupe-sized submunitions that were to be carried by the SS-18 intercontinental ballistic missile. The Soviet Union also considered the following BW agents, falling within the category of incapacitants and developed as weapons for a more tactical role: *Francisella tularensis* (tularemia), *Burkholderia mallei* (glanders, replacing *Brucella* bacteria in the Soviet arsenal), and Venezuelan equine encephalitis (VEE) virus. Also, botulinum toxin was produced in the 1970s, while Marburg virus was introduced as an offensive biological weapon as late as 1989.

The Soviet Union had developed a veritable strategic doctrine concerning the development and use of biological weapons during the cold war. According to Ken Alibek, former Colonel and deputy director of Biopreparat, a cover name for the Russian BW program, biological weapons were not to be used in tactical situations, i.e., at the battlefront, but in a combination of strategic and operational targets, focusing respectively on population centers and enemy logistics. Strategic biological agents were mostly lethal, such as smallpox, anthrax, and plague; operational agents were mostly incapacitating, such as tularemia, glanders, and Venezuelan equine encephalitis. The use of both types of weapons was envisioned on a massive scale, causing huge numbers of casualties and extensive disruption of vital civilian and military activity.

In contrast to the policy held firm during the US BW program, said Alibek, "the Soviets' view [was that] the best biological agents were those for which there was no prevention and no cure."[31] Although the Soviet Union signed the BTWC in 1972, the Soviet Ministry of Defense maintained and even dramatically extended its biological weapons programs. These consisted of several military installations equipped with high incinerator stacks and cold-storage bunkers located in Aksu, Berdsk, Omutminsk, Pokrov, Sverdlovsk (now Yekaterinburg), and Zagorsk (now Sergiev Posad).

The Biopreparat Complex

In 1989, a defector named Vladimir Pasechnik first revealed to the West the extent of the BW program in the Soviet Union. Under the cover of a civilian biopharmaceutical complex called "All-Union Scientific Production Association Biopreparat," an umbrella BW organization was established in 1973 soon after the USSR signed the BTWC. The program consisted of some 20 research, testing, and production facilities located throughout the Soviet Union, employing more than 25,000 people. Taking into consideration all relevant bureaus and institutes involved with Soviet BW research, the total number of personnel may have reached as high as 60,000 in the late 1980s and early 1990s. Biopreparat employed elaborate cover stories, maintaining that certain facilities only manufactured biopesticides (such as *Bacillus thuringiensis*), fertilizers, or vaccines. Biopreparat in fact dealt with both legitimate commercial and military-related activities. Based on the number of medals awarded to Soviet BW researchers, the Soviet Union had probably weaponized plague bacteria successfully in 1978.

Soviet work on plague bacteria was conducted at a military institute in Kirov, research on *Bacillus anthracis* (anthrax) was carried out at Sverdlovsk and Stepnogorsk, while *Francisella tularensis* (tularemia bacteria) was studied at Podolsk. Due to the need for secrecy, a significant effort was made to camouflage the outer appearances of selected BW facilities. The Institute of Applied Microbiology in Obolensk, for example, was designed to appear as a typical hospital (at least from above.) The anthrax-manufacturing facility in Stepnogorsk also utilized special construction methods and configurations to avoid being detected by satellite photography.

BW in Russia Today

In 1992, the then President of Russia, Boris Yeltsin, finally admitted that the source of the Sverdlovsk epidemic had been an accidental release of anthrax spores from a military BW facility. But already by 1986, responding to pressure from the Second Review Conference of the BTWC, the Soviets began to convert its military BW facilities to produce legitimate products for the national civilian economy.

Very little is known of the extent to which the former Soviet BW program continues to operate, since Russian participation in international collaborative efforts to deal with stockpiles of Soviet weapons has dwindled in recent years. While the US, Britain, and Russia have conducted some joint investigations of potential BW production facilities, the situation remains unclear.

The existence of earlier plans by Soviet BW scientists to genetically alter smallpox, as well as evidence from Russian scientific literature suggest that

Russian scientists may have continued work on this virus. This includes the introduction of Ebola and VEE genes into smallpox to create a whole new kind of BW agent, a so-called chimera virus. This would not only resist vaccine or anti-viral treatments, but also have a synergistic effect. In the late 1980s, a strain of a combined *ectromelia* (mousepox, a close relative of smallpox) and VEE genes created symptoms of both diseases in laboratory animals. Ken Alibek is convinced that this BW-related research work still continues in Russia, despite Yeltsin's decree banning all such activity.[32]

CHAPTER 9

Control and Disarmament

Biological weapons have presented unique and vexing challenges to those who have wished to eradicate them. Disarmament has been particularly complicated because nearly all biological weapons technologies and agents have legitimate industrial uses and, consequently, BW programs can easily be hidden from even the most intrusive inspector. The most ambitious international effort to control the proliferation of BW, the Biological and Toxin Weapons Convention of 1972 (the BTWC), has so far been largely unsuccessful. The provisions of the 1993 Chemical Weapons Convention (CWC) are enforced by an international monitoring body, the Organization for the Prohibition of Chemical Weapons (OPCW), based in the Hague, and made up of those nations that have ratified the CWC. For the BTWC, there is no such organization, largely due to the continuing disagreements, even among those who have signed and ratified the convention, about how exactly many of its terms should be interpreted and enforced.

In this chapter, we will chart the development of the BTWC and the bumpy and still-incomplete road to a workable international agreement, which actually began some 75 years ago. We will also take a somewhat closer look at a few fairly recent international BW disputes that have arisen since the convention was first set forth, and try to describe in some detail the challenge facing the BTWC and its signers.

HISTORY

The history of international attempts to ban biological weapons actually begins in 1925, with the Geneva Protocol. Although this international convention, coming as it did after World War I, was mainly in response to the use of chemicals as weapons ("the use in war of asphyxiating, poisonous or other gases..."), it did explicitly address biological agents as well, saying that the parties accept not only CW prohibitions, but agree:

... to extend this prohibition to the use of bacteriological methods of warfare and agree to be bound as between themselves
according to the terms of this declaration. . . .[1]

Following this ban, which entered into force in 1928, there was little open international discussion of biological weapons or their elimination. Beginning in 1959, however, discussions began anew as a small consortium of nations began meeting in Geneva again to discuss disarmament. At meetings that stretched from 1962 to 1968, as part of a group called the Eighteen-Nation Disarmament Committee, the United States and the Soviet Union eventually both proposed disarmament plans that abolished chemical, biological, and nuclear weapons. On November 25, 1969, partly as a result of these meetings, and partly as a result in part of the mounting world-wide consensus against both chemical and biological weapons, President Nixon unilaterally renounced both chemical and biological weapons and declared the end of offensive BW activity in the United States. The logic underlying Nixon's rather sudden move to renounce biological warfare was, of course, complex. Sensing that biological weapons were one area in which the United States could use the US nuclear arsenal to face "asymmetrical" threats—that is, threats by smaller, less well-armed nations—Nixon reasoned that there was, especially in light of their unpopularity, no compelling reason to develop and stockpile biological armaments, particularly considering the negotiations that had been ongoing in Geneva. Though his reasons for taking this action may have been motivated more by considerations of *realpolitik* than by idealism, the net effect was to bring these weapons before a very large group of nations in a very public series of international forums that lasted for almost a decade.

Summary of the BTWC

In 1972, following two years of additional negotiation, the BTWC was complete. The Conference of the Committee on Disarmament, which grew out of the Eighteen-Nation Committee, was ready to submit the text of the document to the United Nations General Assembly. The key provisions of the Convention were as follows:

- *Article I* addressed future BW threats by prohibiting the development or acquisition of BW agents and delivery devices, *except* for use in peaceful activities. As we learned with the CWC, this important exception allows nations, organizations, and other groups to develop a defensive BW capacity and conduct research and development using BW-related pathogens and technology.
- *Article II*, which addresses disarmament more directly, mandated that any

existing BW arsenals in a country had to be destroyed or diverted to peaceful uses within nine months following the date on which the Convention went into force.

- *Article III* prohibits the member nations from helping other states or organizations—directly or indirectly—develop a BW capacity, thus serving as a nonproliferation measure.
- *Article IV* requires that member nations create legislation or other appropriate mechanisms that would prohibit biological weapons activity within their borders, or in any other territories over which they have control.
- *Article V* states that parties to the Convention have to consult and cooperate with one another in the event of a BTWC-related dispute, "within the framework of the United Nations and in accordance with its Charter."
- *Article VI* outlines the steps that must be taken if one member state wishes to accuse another of a BTWC violation—complaints are lodged with the Security Council of the United Nations.
- *Article VII* requires member nations to offer support or assistance if another member country, after appeal to the Security Council, is found to be exposed to danger as the result of a violation of the Convention.
- *Article VIII* reaffirms the BTWC as following the Geneva Protocol of 1925, but only alludes to the prohibition rather than explicitly stating it, which is a particular sticking point for China, Iran, and several others.
- *Article IX* requires member countries to continue in good faith negotiations for the international control of chemical weapons.
- *Article X* protects more peaceful uses of BW technology, emphasizing the benefits of both the biological agents and the associated technology for public health sector and the biotech industry. Article X also reassured member nations that the BTWC would not hinder economic or technological development or violate industry confidentiality by releasing proprietary information.
- *Article XI* allows any member state to propose new amendments to the BTWC.
- *Article XII* states that all of the member states will meet in Geneva in approximately five years to review and refine the Convention. Member nations can request reviews prior to that time, however.
- *Article XIII* asserts that the Convention will continue indefinitely, but individual states have the right to withdraw their support under certain conditions.
- *Article XIV* outlines the procedures for signing, ratifying, and activating the BTWC and designates the US, the UK, and the USSR as Depositary Governments responsible for receiving ratification documents from other states and notifying the UN when the materials are received.

• *Article XV* further requires that these three states have archived copies of the BTWC in five different languages and distribute certified copies of the Convention to all member states.

After much discussion, the Convention on the Prohibition of the Development, Production and Stockpiling of Bacteriological (Biological) and Toxin Weapons Convention and on Their Destruction (this was the new and longer title of the BTWC) was opened for signature on March 26, 1972. It entered into force, meaning that a sufficient number of nations had both signed and ratified it, exactly three years later, on March 26, 1975.

At the time of this writing, 143 nations, or approximately 75 percent of the recognized states world-wide, have signed and ratified the BTWC. And "to the extent that membership is an indicator of success, the world's nations view the BWC as a significant arms control agreement with the potential to enhance international security."[2] But successful, world-wide disarmament cannot be achieved through widely-supported documents alone. Extensive and realistic protocols are needed to ensure that the member states comply with the terms. The BTWC, aside from vague directives given in Articles V, VI, VII, unfortunately lacks a cohesive plan for guaranteeing compliance and building international confidence. The effects of this shortcoming are substantial and come into play in the Review Conferences described below. However, it is important to note that the absence of concrete verification mechanisms was a less pressing issue when the BTWC was drafted in the early 1970s for two reasons. First, the Convention was drafted during the Cold War, a time when intrusive inspections were "politically unacceptable, infeasible or unnecessary."[3] Second, while biological weapons were certainly a source of concern, they were not seen as a substantial military threat. Both ideas were to change dramatically over time as BTWC violations occurred and the biotechnology industry grew.

The Review Conferences

In accordance with Article XII of the BTWC, the participating states agreed to hold a conference within five years of the date on which the convention entered into force, and at any other time when a majority of the States Parties to the convention wish to organize such a conference. Since then, several of these review conferences have been held.

1980

The first Review Conference was spurred in part by the advances in genetic engineering and the increased military interest in the biological sciences around the world.[4] Even in light of these developments, the delegates concluded that that Article I adequately covered the dynamic growth in biotechnologies, and no dramatic changes were required. However, two concurrent controversies cast considerable doubt on the practical effectiveness of the BTWC.

In September 1981, during the second part of the Review, the United States accused the Soviet Union and its clients of using a form of biotoxin against anti-Communist guerrillas in Laos, Cambodia, and Afghanistan. While there is persuasive evidence to show mycotoxins were involved in Southeast Asia in the late 1970s and early 1980s,[5] conclusive evidence has yet to prove Soviet surrogates used trichothecene (T2) mycotoxins as a means of warfare. However, the allegations alone raised serious concerns that the BTWC was in dire need of revision as it lacked a means of addressing these problems.[6]

It was a second incident, however, that caused much more serious concern. In 1979, Dr. Faina Abramova, an experienced pathologist working in the Siberian town of Sverdlovsk, began noticing an unusual and troubling pattern in her autopsies. Looking more closely, Dr. Abramova was able to almost immediately confirm that many of the individuals had died of inhalation anthrax. Her conclusion was confirmed by the tell-tale swelling and hemorrhage of lymphatic tissues in the chest and the striking, red Jello-like appearance of the "cardinal's cap" on the surface of the brain. But her almost on-the-spot and definitive findings were quashed by Soviet authorities, who claimed that the victims had been stricken with gastrointestinal anthrax—a much less lethal form of the disease—after eating contaminated meat.

Before the 1980 meeting, the United States had gone public with its suspicions about the incident. The Carter administration claimed that even though the Soviet Union had signed and ratified the BTWC, production of biological weapons had actually accelerated, and what had happened at Sverdlovsk was more than a case of tainted food.

As the story finally emerged, seven years after the BTWC was promulgated and four years after it was put in force—with the Soviet Union one of its key signatories—a terrible accident occurred at a Soviet BW facility in Siberia. A filter at the Compound 19 facility in Sverdlovsk, where weaponized anthrax was being produced in massive quantities, was accidentally removed, allowing *Bacillus anthracis* spores to become aerosolized, and then vented from a high-level containment facility and into the air outside. The horrified Soviet scientists, as soon as they realized what had happened, raced to replace the missing filter, but it was too late.

In a light wind, the spores drifted south from the facility and wafted toward unsuspecting citizens. In this now decades-old tragedy in which at least 68 people died from inhalation anthrax, we get a glimpse of the intricacies of the BTWC and compliance issues. Was Sverdlovsk a textbook example of a verifiable BTWC violation? Much has been written about the outbreak. And numerous scientists, including microbiologists from several countries, studied the data and eventually—more than a decade after the incident—were able to perform detailed on-site studies.

One of those scientists was Mathew Meselson, a distinguished molecular biologist from Harvard and a former CBW advisor in the Nixon administration. He at first maintained that at least insofar as the Sverdlovsk incident was concerned, the Soviet Union may still have been in technical compliance with the BTWC. Meselson noted, among other things, the limited number of fatalities in a ceramic pipe factory located about 3 kilometers downwind from the release point. At this site, there were 100 or more people who could also have been exposed, but for whatever reason just 2 had become ill. Meselson reasoned that "the weight of spores released as aerosol could have been as little as a few milligrams or as much as nearly a gram."[7] The significance of this finding, if true, is that either the anthrax spores were of very low quality, or the actual amount of anthrax was so tiny that it could be measured in milligrams. This meant that, lacking other evidence, one could plausibly claim that the BW activity at the Sverdlovsk facility was purely defensive, and defensive research into BW agents is permitted under the Convention. As Jeanne Guillemin, also writing about the Sverdlovsk incident, asserted, "the claim that the amount of anthrax released was so great that it exceeded any possible peaceful purpose is not supported by this information. . . . We cannot be sure that the Compound 19 facility in particular was in violation of the [BTWC] treaty."[8] There was, in other words, insufficient information to say with certainty that the deaths were caused by a large release of "high-quality" spores.

But as it turned out, the Sverdlovsk tragedy was a case of offensive BW activity, and it was activity engaged in perilously close to a large civilian population. Had the accidental release taken place during the day rather than at night, or had the wind been blowing north instead of south, the result would have been a catastrophe—certainly with hundreds and possibly with thousands of deaths.

Over the years, bits of information about the incident have emerged. The former Soviet "bioweaponeer" Ken Alibek has heard that hundreds of people in fact died in the Sverdlovsk release, not the 68 now confirmed dead by the Soviets, and the source of anthrax likely involved anywhere from grams to hundreds of grams of material. While Alibek's figures may not have the same air of precision as those in Meselson's study, his estimates seem much more plausible. And Meselson himself, in an interview on the PBS television show *Frontline* after he had been given an opportunity to work with on-site epidemiological data, came to believe that the evidence showed unequivocally that accident spread anthrax spores through the air, and in quantity:

> . . . It wasn't until we took a team of independent scientists . . .
> and went to Sverdlovsk and we interviewed 43 families and asked
> them where the person who they had lost . . . worked, where
> they were in the daytime. Then when we plotted those locations
> on a map, they all fell on a very narrow straight line, a very
> narrow zone, and it turned out that the wind was blowing in that
> ᵥery direction on one of the days just before the first cases. [The]
> line went down 50 kilometers south of the military facility along
> which there were villages where animals had died of anthrax
> and one end of this zone was the military facility. That answered
> it unequivocally.[9]

But if even someone as responsible as Meselson can have doubts on the Sverdlovsk incident as a clear BTWC violation, how can the BTWC hope to be able to separate real infractions from false allegations? With no lack of its own lawyers, even the United States could conceivably cook up similar and contrived justifications when challenged with a BTWC violation.

1986

Still reeling in the throws of the Sverdlovsk incident, the BTWC reconvened in 1986. In order to increase the level of trust among signatories and improve transparency, four important confidence building measures (CBMs) were established. The first required that member nations submit annual reports on any high containment facilities designed for work on dangerous biological materials. Second, member nations were required to notify the BTWC of outbreaks of any unusual diseases potentially caused by BW. Third, the BTWC encouraged the publication of any BW-related research. Finally, member nations were encouraged to promote more exchanges among scientists involved in related research.[10]

The overall response to these CBMs has been tepid, with only 30 to 40 countries reporting regularly on an annual basis since the 1986 Review.[11] Most developing nations have either failed to submit the required documentation or have handed over paperwork which was incomplete.[12]

1991

At the 1991 Review, considerations for strengthening the BTWC were greatly inspired by the just concluded Gulf War (1990–1), and many new or expanded Articles were proposed. For example, some delegates wanted to extend Article 1 to cover BW agents against plants and animals. Additionally, it was decided that member nations would provide data on their national BW defense programs and facilities. These declarations would also include information on any offensive or defensive biological programs initiated since January 1, 1946, and details on vaccine production facilities.

The Ad Hoc Group of Governmental Experts to Identify and Examine Potential Verification Measures from a Scientific and Technical Standpoint (VEREX) was subsequently established by the Third Review Conference, which met in September 1991. The group's final report, issued in September 1993, identified 21 potential verification measures that, if implemented in appropriate combinations, would provide sufficient confidence that prohibited activities were not occurring. Additional meetings of the VEREX group were held in Geneva, between March 1992 and September 1993, resulting in a final report to the BTWC States Parties. While VEREX was able to conclude that at least a combination of the measures listed above was promising, it did recognize that the dual-use nature of BW related technology would make verification problematic.[13]

1997

At a "Formal Consultative Meeting" in 1997, another controversy erupted—this time involving accusations against the United States. At the meeting, the Cuban government claimed the US had released an aggressive, burrowing and quite voracious insect named *Thrips palmi* on the island of Cuba.

The Cubans claimed that, on October 21, 1996, a US aircraft on a flight plan approved by both American and Cuban governments flew over the island nation's Giron corridor. Having observed a Cuban aircraft in adjacent air space, the US pilot—as the US claimed is customary in such situations—released a smoke-generated signal as a proximity warning, ensuring that both planes were able to safely make visual contact. This seemingly unimportant event was used by the Cuban authorities to charge the United States with the deliberate dissemination of the *Thrips palmi* insect in a plot to destroy crops on the western part of the island. First detected in December 1996, the outbreak of the insects would have been consistent, at least according to the Cuban allegations, with the "mysterious" smoke issuing from the US aircraft two months before.

Cuba took their case before the international community, invoking Article V of the BTWC that allowed for special consultation among States Parties to the treaty. With the United Kingdom serving as the chair of the Formal Consultative Meeting late in the summer of 1997, three sessions were held during which Cuba and the United States were allowed to make their cases.[14]

Raymond Zilinskas has analyzed the *Thrips palmi* and other cases at length, and we only need summarize the basic arguments here.[15] Cuba told the representatives that, due to the exact timing of the US aircraft overflight, its suspicious emissions, the outbreak in West Cuban farms with *Thrips palmi*, and with no other plausible source of the infestation, that clearly the United States had engaged in a form of BW. As for the US retort that the aircraft in question was only displaying a smoke signal to warn approaching aircraft, Cuban delegates said that, on the contrary, American aircraft were not typically equipped with such devices, and the signaling protocol was anything but ordinary. In response, the US representative noted that it was impossible to prove association with cause. That is, the presence of an aircraft and a plague of insects by themselves were flimsy evidence for a significant allegation of BW. More importantly, there were scientifically proven cases of similar, naturally-caused infestations emanating from airborne transmission in the region, as well as trade in agriculture that could have brought in the unwanted visitors. The US delegation further emphasized the use of warning smokes was a normal practice.

It is instructive to note that all but two of the nations who heard these arguments found no substance to the Cuban charges. The Hungarian representative wrote, in no uncertain language, that he saw "no link between the overflight and the infestation."[16] Denmark noted that during the sessions, the US government "convincingly demonstrated that the occurrence of *Thrips palmi* in the Matanzas province of Cuba ... could have resulted [from] a number of causes, including natural phenomena as well as the normal movement of trade and goods."[17] Both the Netherlands and Germany concurred, the latter pointing out that "insects such as *Thrips palmi* couldn't be dispersed from an aircraft as dry substance."[18]

However two States Parties, China and North Korea, disagreed with the majority, to differing degrees. China stated only that its experts found it "hard to draw conclusions" on, among other details, whether or not the United States actually disseminated the insects over Cuban airspace.[19] North Korea was decidedly less guarded, finding it "regrettable that the incident of spraying of biological substances by the United States against Cuba has taken place."[20] Notably, the two countries that continue to foster and promote the accusation that the US employed biological weapons in the Korean War also found Cuba's recent allegations possible (in the case of China) or persuasive (in the case of North Korea).

As Milton Leitenberg has emphasized, extraordinary allegations require substantive evidence.[21] Of course, one requirement for securing substantive evidence has to be the ability of the investigator to seek out that evidence. And as mentioned earlier, the BTWC, which preceded the CWC by twenty years, lacks that later convention's "teeth" in terms of verification and inspection.

THE BTWC TODAY

Even as we focus on the government's role in disarmament, it is important to remember the other interested parties. The biotech industry—including those involved in pharmaceuticals, food products, beverages, etc.—sees itself as a responsible consortium of developers and manufacturers, willing to provide relevant information and expertise to an international secretariat monitoring compliance with the BTWC. But industry representatives also sound caution-ary warnings about the degree of certainty that can be achieved under such a convention and its protocols, and the need to place the protocol within proper scientific perspective.[22]

Verification and monitoring protocols of the BTWC could follow in the footsteps of the recently ratified Chemical Weapons Convention. The CWC was, by and large, supported by the chemical industry worldwide, particularly by the Chemical Manufacturers Association (CMA). However, there are important distinctions between those measures intended to control both chemical and biological weapons, especially from the standpoint of the biotech industry. With respect to the BTWC and verification measures, the biotechnology industry points out that a large component of its equipment is "dual use" which could, in theory, be used in the manufacture of biological weapons with little or no modification. Furthermore, technologies made sus-picious under the BTWC may be found in academic labs as well as food and beverage fermentation plants and fuel processing facilities. Also, modern clean-in-place technology allows for production pathways to be cleared within as little time as one or two hours, allowing BW producers to quickly mask their activities.[23]

In contrast to the success of the CWC in gaining acceptance from chemi-cal manufacturers, a different picture emerges as the biotech (especially the pharmaceutical) industry evaluates future BTWC verification protocols. While fully in support of the general goals expressed in the BTWC, the organization that represents many US and international biotech firms, the Pharmaceutical Research and Manufacturers of America (PhRMA), is partic-ularly concerned about intrusive inspections in this rapidly-growing industry. Unlike the many chemical manufacturers that are generally comfortable with the intrusive nature of CWC verification procedures, the biotech industry and their organizations (such as the PhRMA) are much less sanguine about rou-tine BTWC inspections. While US-based research in chemistry secured more than half of the world's patents issued in 1997, the biotech field is represented even more by American firms, receiving over 65 percent of all biotechnology patents that same year.[24] Genetic engineering and its products, for example, are arguably the most important features of the new biotech industry, and it is for good reason that trade secrets concerning this type of technology are jeal-ously guarded.[25]

In the case of CWC verification methods, raw materials can indicate that chemical weapons are being produced and therefore precursor control and

Dual-use biological processing equipment: fermenters at an Iraqi palm date production facility. UNSCOM inspectors examined these for their potential for BW agent production, September–October 1991. (United Nations, Photograph by H. Arvidsson.)

managed access may be adequate to monitor the chemical industry as a whole. However, in the case of BW agent production, aside from the microorganism or toxin involved, there is no component or raw material that is necessarily unique between legitimate commercial production (vaccines, for example) or the production of botulinum toxin for a weapon. In the biotechnology field, raw materials are less important than the proprietary organism and/or finely tuned processes, the latter technologies requiring years of research and large amounts of investment capital.[26]

THE FUTURE OF THE BTWC

In July 2001, bringing both doubt and some recrimination from the arms control community, the United States rejected the latest version of a proposed BTWC protocol. US Ambassador Donald Mahley listed the main concerns the United States had over the BTWC Protocol text, including its inability to detect or deter countries from acquiring biological weapons and the lack of protections provided for commercial business information. Notably, Ambassador Mahley also said that, as currently conceived, the Protocol would not have protected US biological defense programs, while not doing enough to ensure that actual offensive work was not being done elsewhere. Finally, Mahley noted that some countries were using the BTWC negotiations as a pretext to undermine other export control regimes they did not like. Once it became apparent that the US had rejected the latest version of the protocol, other countries decided it was not worth pursuing negotiations further. While the US rebuff of the recent verification and compliance scheme does not mean America has withdrawn from the treaty, it casts a shadow on the viability of a future diplomatic consensus on the BTWC and its implementation.

CHAPTER 10

Vaccination and Biological Warfare

Whether it was the spread of plague following Mongol invasions, or the near eradication of Native Americans due to disease brought in from the Old World, disease has played a significant role in strategy and geopolitics. Of the many lives lost due to the violence of war throughout history, many more have succumbed to disease during battle rather than actual combat. Typhus (spread by lice), plague (via fleas and infected rats), and dysentery have caused enormous numbers of deaths in major wars as far back as one can go. For every one British soldier killed by Russian rifle or artillery, for example, at least ten died from dysentery during the Crimean War (1854–56). Ten years later during the American Civil War, infectious disease also took a disproportionate toll.

Deaths of Union soldiers due to disease versus wounds, American Civil War[1]

Cause of death	All soldiers
Killed in action	67,058
Died of wounds	43,012
Died of disease	224,586

The onset of the devastating Spanish Influenza epidemic of 1918 that caused thousands of casualties on both sides during World War I may have been partly responsible for Germany capitulating in July of that same year. (For a short period of time after the war, some even suspected Germany of deliberately unleashing the virus that killed some 20–50 million people worldwide.)

DISEASE AS DETERRENCE

Having already been invaded by the Mongols in the mid-thirteenth century, China was for a long time since under constant threat of attack from "northern barbarians." But smallpox, being relatively common among the southern Han people, formed a natural, albeit temporary, barrier to invasion by the northern peoples of Manchuria. During the Ming dynasty (1368–1644), the Great Wall was somewhat effective in repelling the southward advances of Manchurians. But there was also the terrible fear of contracting smallpox that caused the Manchus, along with other northern nomadic tribes, such as the Liao and Khitan Tartars, to hesitate in their adventurism:

> ...Whenever the northern peoples, such as the Mongols,
> Manchus, the Liao (Khitan Tartars), Tungus, etc., invaded the
> Chinese Central Plains, they often would contract smallpox from
> contact with the Han people, and as a consequence would not
> dare attack south of the Great Wall. . . .[2]

If military operations had to be conducted in the central plains, northern tribe commanders only chose to send those Manchu soldiers who had previously been infected with smallpox. As did Thucyidides 2000 years before,[3] it was known in China that if one survived smallpox infection lifelong immunity was conferred. As fate would have it, 17 years after the Manchurians finally conquered China, their Emperor Fulin (Shizu) died from smallpox in 1661.

When it came to managing disease in the military context, dramatic improvements were made by the mid-twentieth century, but even then infectious disease claimed thousands of lives. Despite efforts to immunize against disease—including the use of typhus and cholera vaccines for Germany's *Afrika Korps* during World War II—sickness still claimed a large percentage of the *Wehrmacht's* most elite troops. During his campaigns in Africa, Field Marshal Erwin Rommel, "the Desert Fox," himself made repeated trips back to Berlin due to infectious disease.[4] As demonstrated by newsreels showing him often with a handkerchief to his nose, Rommel suffered from nasal diphtheria. Reported Rommel's medical advisor in August 1942:

> ... Field Marshal Rommel suffering from chronic stomach and
> intestinal catarrh, nasal diphtheria, and considerable circulation
> trouble. He is not in a fit condition to command the forthcoming
> offensive. . . .[5]

Rommel's bout with infectious disease may have influenced the very outcome of the African campaign in World War II.

Smallpox, Variolation, and the First Vaccines

Smallpox virus has been a centuries-old threat to military forces, and there was little one could do to avoid smallpox until at least A.D. 1000, when the Han Chinese probably first learned of variolation. In ancient and imperial

China, various miasmas were credited with having decimated large armies and thereby influencing large military engagements. But even before the germ-induced disease concept had been postulated, and long before science had even a basic understanding of the virus responsible, it was already known that through the controlled exposure to smallpox one could avoid contracting the full-blown disease. This technique, known as "variolation," was employed to inoculate people against smallpox for many centuries.

The practice of variolation is said to have been used in Africa in ancient times, and may have been practiced by Buddhist monks at the Mount E' Mei temple in Sichuan province sometime during the Renzong dynasty (1022–63). These monks were said to have originally learned the procedure from the Tibetans, who had earlier been taught by Indians. By the 1500s, variolation was definitively mentioned in Chinese medical texts.

The inoculation process was accomplished in two main ways. In the first, material likely to be related to the disease, such as that from a recently raised smallpox lesion, was administered to an immunologically naïve person by scarification (scratching the skin). Scarification was the standard technique for variolation in Turkey and Europe, but was decidedly different from variolation practiced in China. While the former introduced the inoculum through broken skin, the Chinese method generally consisted of using dried smallpox scabs, pulverizing them into a powder, then blowing the infectious material into nasal passages. As one might expect, either procedure was extremely dangerous for there was little guarantee that full-blown smallpox would not result. But in most cases (approximately 99 percent),[6] variolation caused only mild disease and protected the individual from future smallpox exposure. The reasons why this method was effective, or even more so why it did not result in more cases of actual smallpox, are still obscure. It is possible that the type of inoculation, and the fact that much of the virus was probably dead before scarification took place (but still able to form antibodies), may have played important roles.

The first mention of variolation used in a military context was probably that of the Manchurian Kangxi emperor, whose brother Shizu had himself died of smallpox. In 1661, after many years of opposition by his predecessors to institute such a policy, the Kangxi emperor finally decreed that his military carry out variolation for the troops. This would predate by more than a hundred years General George Washington's order that Continental Army soldiers undergo a similar procedure.

The first reference to the Western practice of variolation actually starts in the early 1700s. Giacomo Pylarino, a Greek who had been in Constantinople during a serious outbreak of smallpox, witnessed a woman who excised pustule matter from a patient for "transference." This procedure employed the grafting of smallpox material to another person and was later called "inocula-tion" (coined by Emanuel Timone in 1714.)[7] When posted with her diplomat husband in Constantinople, Lady Mary Wortly Montague observed close at hand eighteenth-century Turkish society. Having contracted a mild case of smallpox herself as a young woman, she was fascinated to learn of a local prac-tice that exposed children to the virus in order to protect them from serious infection later on in life. In this already age-old practice in Turkey, old women found mild cases of smallpox victims, took the exudate from their smallpox lesions, and collected them in nutshells. The elder nurses went from house to house, visiting children and others who were to be inoculated. A large needle was used to make a small tear on the skin near a vein on the arm or leg, the needle then covered with the smallpox lesion exudate, and then placed on the newly introduced wound. Finally, the scratch was bandaged with a shard of the original nutshell. After about eight days, the children would contract a fever, be put into bed for two or three days, "and in eight days' time," wrote Lady Mary, they were "as well as before their illness."[8] Its success rate in preventing serious cases of smallpox was quite evident to other European contemporaries who knew of the practice.

The first known variolation in England proper was that of Lady Mary's son, Edward Wortley Montague, in 1718,[9] and this form of vaccination quickly became widespread throughout Europe. At about the same time, the Bostonian Cotton Mather, a theologian and self-taught scientist, helped a local physician (Zabdiel Boylston) to inoculate 286 people in the middle of a small-pox outbreak. Mather had learned much earlier of variolation from discussions with his black servant and subsequent reading of its use in Turkey. Using a sta-tistical analysis of the Bostonian experience, comparing those who had been vaccinated to those who were untreated, Mather recognized the benefits of variolation in spite of the well-recognized risks.[10]

The British Army also adopted the practice of variolation in the early 1700s. Because many American colonists had themselves previously served in the British Army, this knowledge was made widespread by the time of the Revolution. General George Washington, for one, had more than passing knowledge of smallpox, having suffered himself from it during a brief stay in Barbados as a young man. When it came time for the Revolution in 1776, smallpox was quite common in America, especially among Continental Army soldiers. Eventually, General Washington instituted mandatory variolation for all soldiers in the Continental Army. Due to the dangers inherent in such a procedure, this policy was controversial even then, and inoculations had to be carried out in secret.

In April 1776, John Morgan, who was Physician-in-Chief to the American Army, urged that widespread variolation be conducted for the troops, and in his January 6, 1777, letter to William Shippen, Jr., George Washington made it official:

> ... Finding the small pox to be spreading much and fearing that no precaution can prevent it from running thro' the whole of our Army, I have determined that the Troops shall be inoculated. This Expedient may be attended with some inconveniences and some disadvantages, but yet I trust, in its consequences will have the most happy effects. ... I have directed the Doctr. [Nathaniel] Bond, to prepare immediately for the inoculating this Quarter, keeping the matter as secret as possible, and request, that you will without delay inoculate all the Continental Troops that are in Philadelphia and those that shall come in, as fast as they arrive. ... I would fain hope that they will soon be fit for duty, and that in a short space of time we shall have an Army not subject to this, the greatest of all calamities that can befall it, when taken in the natural way. ... [11]

As variolation was routinized for the troops, it was found that mortality due to the inoculation procedure was approximately 1 in 300, compared to 16 percent for naturally acquired smallpox. Losses due to this disease dropped dramatically for the American Army thereafter, no doubt influencing the outcome of the Revolutionary War.

Jenner's Vaccinia

Despite the clear effectiveness of variolation it remained a very dangerous method of inoculation. This would change dramatically upon the introduction of Jenner's vaccine in the late eighteenth century.

In the 1790s, the English ornithologist Edward Jenner decided to take on a career in medicine, although it is unclear as to where he obtained his credentials for such a venture. During rounds in the countryside of Gloucestershire, Jenner discovered that cattle farmers often had no reaction to variolation, nor did others who routinely worked around cattle. Jenner had known of the Old Wives' tale (which turned out to be true) that milk maidens' skin was fair in complexion because these women were immune to smallpox. As postulated at the time, this happy consequence was due to the protection that an infection with cowpox (a rather innocuous disease) afforded to humans. (According to some historians, the Chinese were among the first to understand that inoculation with vaccinia virus could also prevent smallpox.[12]) In 1796, Jenner finally put this theory to the test by inoculating an eight-year-old boy with cowpox and then trying to infect the child with smallpox—eventually more than twenty times—but the boy never became ill.

Edward Jenner's use of the inoculum *Variolae vaccinae* revolutionized prophylaxis against smallpox, ending the risky variolation procedure with the more virulent and disease-causing inoculum. In fact, the very word "vaccine" comes from the Latin for cow (*vacca*), as Jenner's technique utilized the live cowpox virus. (Or so it was believed. Vaccinia virus may have been a horsepox type that has since become extinct.) Having learned of this effective technique, President Thomas Jefferson (no mean naturalist himself) not only inoculated his family three years later with this new vaccine, but he was also responsible for instituting nationwide vaccinations by lobbying Congress to pass the "Act to Encourage Vaccination" in February 1813.

Vaccinia, of course, was basically the same type of virus that eventually eradicated smallpox in the late twentieth century. But in the 1800s, vaccination was by no means universally adopted or even politically favorable. The over-population doomsayer and political economist Thomas Malthus (1766–1834) warned that ". . . if the introduction of the cow-pox should extirpate the small-pox, and yet the number of marriages continues, we shall find a very perceptible difference in the increased mortality of other diseases."[13] Other arguments from anti-vaccination quarters, some widely held, ranged from the notion that the cowpox virus would make people's faces turn bovine in character, to the more reasonable caution that these early vaccine preparations might be dangerously impure. The industrial mogul and president of the Anti-Vaccination League of America, John Pitcairn of Pennsylvania, was the patron of the US movement against immunizations. His rhetoric inspired Charles M. Higgins to write a self-published manifesto in 1920, *Horrors of Vaccination Exposed and Illustrated: Petition to the President to Abolish Compulsory Vaccination in the Army and Navy.* The latter's view was that "medical compulsion, like religious compulsion, is Un-American and must be abolished."[14] Allan Chase, who wrote a definitive history of vaccinations against disease throughout the last two hundred years, described such objectors to immunization in this way:

> These otherwise highly competent laymen had very different reasons for opposing vaccination, but they shared two beliefs. They were convinced
>
> (1) that they knew more about infection and immunity than any doctors and life scientists who had ever lived, and
>
> (2) that Jenner's vaccine could only cause harm to and kill people, and had never prevented a single case of smallpox.[15]

In the nineteenth century, European parliaments balked at the thought of vaccinating the wider civilian populations by governmental fiat, but their own militaries did not hesitate to protect their troops by immunizing them against smallpox. Napoleon, although an arch enemy of England at the time, nonetheless instituted an honorary medallion in the name of Jenner because of his *vaccinia*, and by 1805, the entire French Army had been inoculated against smallpox. In later years, however, while ostensibly dedicated to protecting her army against smallpox, the French military's order to vaccinate was honored more in the breach than the observance.

The aftermath of the Franco-Prussian war (1870–1) demonstrated not only the devastating results of this uneven immunization, but also raised suspicions that France had deliberately used smallpox as a weapon against Prussia. Unlike the French, Prussia, having strictly adhered to a vaccination regimen for its soldiers, suffered relatively few casualties. France, on the other hand, tallied more than 30 times as many smallpox cases among its troops, including 23,470 deaths from this disease during the first half of the war. Prussian military commanders and politicians contended that this wave of smallpox was no accident, but in fact a deliberate plot hatched by the French to spread the disease throughout the Prussian military and populace. Having observed the many captured French soldiers that were infected with variola, in the minds of the Prussian government and within its own military this raised the suspicion of biological warfare. Finally, Prussia's own scientific community was able to present a more convincing explanation: France simply had not adequately vaccinated her soldiers. As a consequence of the pandemic Prussia quickly mandated vaccinations for German youngsters between 2 and 12 years of age, but otherwise did not take any drastic military countermeasures.

Typhus and DDT

According to Ken Alibek, by the 1930s the Soviets had produced both liquid and dried forms of typhus for use in biological weapons. However, there is no evidence that the Soviet Union ever deployed its typhus weapon against Germany. Ironically, it was the Soviet Union that suspected Germany of deliberately infecting Red Army troops with the same pathogen. According to one Soviet retrospective on the Battle of Stalingrad, not only did the retreating German Army leave unsanitary conditions behind, but furthermore,

> . . . [T]he fascists imposed distinctive, epidemiological diversities aimed at injuring our troops: they threw across the front line the lice-ridden victims of spotted fever and prior to their back off dissolved the camps of war prisoners and civilian population infected by spotted fever.[16]

According to a Soviet history of disease during the Battle of Stalingrad, the number of German casualties due to typhus nearly equaled battlefield injuries.

Although an early typhus vaccine was used in World War II for European troops and American prisoners of war in Germany, its real efficacy has always been in doubt. By far the greatest factor in the reduction of typhus was the introduction of DDT to kill its main vector, lice. The United States Typhus Commission, formed in December 1942, was established in the War Department as a military/civilian organization. In early 1943, Merck & Company produced about 500 gallons of DDT, which was immediately sent to Naples, Italy. This proved to be a successful operation and an effective demonstration of how the application of DDT could halt a fast-growing typhus epidemic.

Following war's end, a significant problem was encountered when it came time for the Allies to process thousands of German prisoners of war, many of whom were infected or were vulnerable to infection by typhus. Few German soldiers in that theater had been vaccinated for typhus, and many were ridden with lice. According to the official US military history of medicine in World War II:

> . . . Under the direction of Colonel [John E.] Gordon, and in
> part, the United States of America Typhus Commission, through
> the awareness of medical officers, and by the abundant use of
> DDT insecticide powder, typhus control was so intelligently and
> effectively carried out that the disease, which might have been
> catastrophic, was of minor significance among German prisoners
> of war in the European theater. . . .[17]

Since World War II, DDT has saved untold lives, not just from the scourge of typhus but from many other arthropod-borne diseases. According to a 1970 National Academy of Sciences report:

> . . . To only a few chemicals does man owe as great a debt as to
> DDT. . . . In little more than two decades, DDT has prevented
> 500 million human deaths, due to malaria, that otherwise would
> have been inevitable. . . .[18]

MODERN MILITARY VACCINATIONS

If disease influenced the conduct and disposition of warfare, then it was often the exigencies of war that determined the course of modern vaccine development. Despite the scientific advancements in the nineteenth century, which promised to dramatically lower childhood mortality from now preventable diseases, it was the military demands for *adult* immunoprophylaxis that drove its early applications. It is worth quoting Allan Chase's observations on the earliest developments in immunology with regard to infectious disease:

> . . . During the four years of [World War I], and the two or three
> years of acute disruption and reconstruction which followed it,
> military and military-linked priorities alone determined who
> would benefit the greatest, or eve at all, from the most effective of
> the vaccines and antitoxins developed between 1880 and 1914. . . .[19]

Although not all of the first developments in immunization were successful (some in fact did more harm than good), by 1914 effective and safe vaccines were available for smallpox (vaccinia), rabies, anthrax, and whooping cough (pertussis toxoid), as well as passive protection against tetanus and diphtheria. The latter two developments, particularly the use of tetanus antitoxin, no doubt saved countless lives during World War I.

Typhoid and the Boer War (1899–1902)

By the 1880s, it was learned that a killed typhoid culture could provide an effective vaccine, and this advance led to dramatic improvements not only in terms of public health for armies but general hygiene as well. The Boer War of 1899 saw an effort toward the mass immunization against typhoid for British troops being sent to South Africa, but this initiative quickly became embroiled in a contentious issue over voluntary versus mandatory inoculations. As soon as the suggestion was made, a rumor had sprung up concerning the real or imagined risks from the typhoid vaccine, scuttling the original plan for compulsory immunization. As a result, Almoth Wright was allowed only to use volunteers from the British military for his typhoid vaccination initiative. Even so, the reception to even this modest venture was so hostile that large amounts of Wright's vaccine were pushed overboard while in transit through the port at Southampton. Although 14,000 of its soldiers were eventually immunized, typhoid fever wrought havoc in the British Army, affecting 58,000 and killing some 9000. Debate still continued in England over the efficacy of the typhoid vaccine thereafter, and typhoid inoculations in the British military remained strictly voluntary at the outbreak of World War I.

Although rare in developed countries, typhoid is still a significant problem in poorer nations around the globe. In the United States military, typhoid vaccinations have been required of Navy and Marine Corps forces on alert and all Special Forces, as well as others at risk of acquiring the disease when deployed outside the continental United States.

Yellow Fever as a BW Threat

Although he was himself skeptical that germ warfare would ever materialize as a threat, the British scientist John Burdon Sanderson Haldane first warned of the yellow fever threat as biological weapon in 1938. Following reports of attempts—all of which were unsuccessful—by the Japanese Army to acquire yellow fever virus in 1939, the threat to American troops in the Pacific from this potential BW agent spurred a crash effort in the United States to produce a vaccine. Unfortunately, when it was administered in 1942 to thousands of

American soldiers, a large portion of the human sera-derived, yellow fever vaccine was contaminated with hepatitus B virus. Even when it was suspected that the vaccine was the source of jaundice, eventually affecting at least 50,000 US soldiers, the perceived BW threat trumped the issue of safety and the vaccinations continued.

The most recent estimate, based on the 427,000 doses used from contaminated vaccine lots, is that some 330,000 US servicemen may have been exposed to hepatitus B from the yellow fever vaccine. While one might have expected increased rate of deaths due to cirrhosis or liver cancer, a preliminary analysis of data from death certificates did not show a significant increase in mortality among those infected with hepatitus B during the war.

Japanese Army units, including the notorious Unit 731 led by the war criminal Shiro Ishii, certainly looked upon yellow fever as a possible BW agent. But the only known attempt to use BW against Allied forces was a commando raid on Saipan by the Japanese. Here, plague-infected fleas were to be disseminated on a US air field in Saipan, but the boat carrying Ishii's assault team was sunk by a US submarine, leaving only one survivor.

Japanese B Encephalitis and the War in the Pacific

In 1942, Albert Sabin reported that the United States had developed a lyophilized vaccine for Japanese B encephalitis.[20] Following the battle for Okinawa in 1945, Japanese B encephalitis was found to be endemic among the island natives, with many US troops also coming down with this infection. Making matters worse still, partly as a result of the difficult terrain, suppression of the mosquito vector using DDT in the north could not be accomplished as quickly as it was in the southern Okinawa. While the situation seemed to call for a massive immunization program, the aforementioned vaccine—made from formaldehyde-killed virus from lyophilized mouse brain tissue—could not be produced in large enough quantity. In response to the military requirements, a relatively crude Japanese B encephalitis vaccine prepared from mouse brain was eventually produced by US commercial firms. To ascertain the minimum requirements for adequate inoculation, this vaccine was first tested on 35 people.[21] In 1945, about 60,000 to 70,000 US military personnel in Okinawa were given the vaccine, with no evidence of adverse reactions among the more than 53,000 that received the two doses. By the following year, an additional 250,000 individuals stationed on the island were also immunized.[22] It is still not known precisely how effective this vaccine was in preventing encephalitic infections among US soldiers.

Botulinum and D-Day

When it came time for the massive invasion of Normandy against the *Wehrmacht*, the United States, Great Britain, and Canada had already considered the possible threat from German biological weapons. A year before D-Day, the American and Canadian intelligence services had reason to believe that Germany would use botulinum toxin against the Allies, possibly loaded on V-2 rockets, or used in some other fashion against the beach landing forces.

Relying primarily upon the information provided by a German refugee scientist, Helmuth Simons, both Canada and the United States prepared large amounts of botulinum toxoid as a vaccine for the landing troops. Canada produced at least 25,000 doses of the toxoid, while the United States made enough (over 1 million units) to inoculate at least 300,000 American troops if necessary. The latter was a toxoid vaccine for type A botulism. (Botulinum type B toxoids were also under production but these never made it to Europe.) Due to the danger involved, US and Canadian workers involved in the manufacture and handling of the toxic botulinum extracts were required to be vaccinated for botulinum themselves. And as is the case today, there were some of these personnel who steadfastly refused to take inoculations to protect against accidental botulinum poisoning. Said the official history, "A few individuals refused to undergo immunization, and when explanations of its importance were of no avail, they were assigned to work where they would not be exposed to pathogenic agents."[23]

Recent research by John Bryden and others has uncovered why, despite all of the great effort involved in preparing botulinum toxoid vaccines for the D-Day troops, none of these prophylactic preparations were ever used. Because British intelligence had broken Germany's most secure code, the Enigma cipher, the higher echelon of Allied commanders was reasonably secure in the knowledge that the *Wehrmacht* had no immediate plans to use chemical or biological weapons. However, because the British Ultra decrypts were so strictly classified, security demanded that only the most highly ranked of the Allied leaders had "a need to know." This compartmentalization meant that the scientists in subordinate commands continued with the production of botulinum toxoid unabated. But when it finally came time for the Allied commanders to make a decision, it had already been made not to go forward with the immunizations.

Plague and the Vietnam War (1965–1975)

In 1894, Shibasaburo Kitasato (one of Robert Koch's protégés) and Alexandre Émile John Yersin each traveled to Hong Kong to investigate the outbreak of plague in southern China. (While the name *Yersinia pestis* has been given in

Yersin's honor, it is nearly certain that both he and Kitasato recognized the Gram-negative bacterium at about the same time). In 1895, Yersin established a second Pasteur Institute in Nha Trang, Vietnam, and founded the Medical School of Hanoi. Yersin spent the rest of his life in Vietnam, working primarily in the field of agricultural technology. Over a hundred years later, the Vietnamese continue to commemorate Yersin's contributions, including naming street signs after him and issuing a postage stamp in his honor.

During the Vietnam War, the US Department of Defense (DOD) was well aware that plague was endemic in southeast Asia and therefore undertook a vaccination program for personnel serving in country. Considering the hundreds of thousands of American troops that were sent to Vietnam, the plague vaccine was an astounding success. Combined with the use of pesticide and a better knowledge of plague etiology, the vaccine was so effective that only eight US soldiers were infected with *Yersinia pestis* during America's military involvement in Vietnam.[24]

Botulinum Toxoid and the Gulf War (1991)

During Operations Desert Shield and Desert Storm of the Gulf War, 150,000 US military personnel were administered the anthrax vaccine. But due to limited quantities on hand, fewer of these (about 8000) were able to receive a botulinum toxoid preparation in the event Saddam Hussein's army would use botulinum toxin. This vaccine was prepared from *Clostridium botulinum* cultures (serotypes A through E) and had been previously used for hundreds of US Army researchers since the 1950s. Although it was not certain if such prophylaxis would be effective in the event of exposure to aerosolized botulinum toxin, a week into the air campaign against Iraq, the US Food and Drug Administration finally approved the safety of the vaccine.

During the Gulf War, a moral dilemma faced US and Coalition military commanders: With a limited supply of both botulinum toxoid and anthrax vaccines, who should receive them? Some commanders clearly regretted that only partial vaccinations were performed, and would preferred to have the risk equally parceled out, rather than only treating an arbitrarily chosen group. It was finally decided that some was better than none, and those troops and personnel stationed in areas more likely to be under a BW threat were given priority for inoculation.

For passive protection against botulism, there is one especially interesting immunoglobulin form of antitoxin that has shown promising results. It is derived from horse sera. In fact, following minute and gradual administration of the toxin for many years in an old army horse, First Flight, over a thousand liters of blood containing antibodies to botulinum have been drawn from this single animal.

Vaccinating for Anthrax in the Twenty-First Century

In light of the extent of Iraq's biological weapons program, the possible threat from other hostile nations that may possess biological warfare agents, and the possible use of biological warfare (BW) agents by terrorists, the United States has instituted mandatory anthrax vaccinations for all US military personnel. The program was formally instituted on December 15, 1997, and since then over 425,000 persons in the uniformed services have received at least one of the six injection series for anthrax, totaling 1,620,793 doses.[25] Current plans are to have all US military personnel (approximately 2.4 million) receive the vaccine by the year 2005.

While representing only a small percentage of all US Department of Defense personnel, there is nonetheless a sizeable number of men and women in the US military who are resisting the required immunizations, citing primarily safety concerns over the Anthrax Vaccine Adsorbed (AVA) vaccine. By refusing to be vaccinated against anthrax, in continuance with military code going back to World War I regarding mandatory vaccinations, US military personnel are subject to disciplinary action for disobeying orders. Over 350 so far have said they will not submit to vaccination for anthrax, and have been threatened with punishments that include courts-martial. As of March 2000, about 700 members of the US Air Force reserve are resisting the required inoculations, preferring rather to resign or be transferred. Additionally, an estimated 100 civilian contractors for the DOD have also refused the required vaccine, and these individuals are also subject to disciplinary action.

Historical Development

A Soviet-made, live attenuated (non-encapsulated) strain of anthrax, administered in 1943, is probably the first recorded anthrax vaccine made for humans. At about the same time, one of the first orders of business in the United States' biological warfare program was also the development of anthrax vaccine. In one of the first efforts, 205 personnel were administered a vaccine made of killed *Bacillus anthracis* bacteria (the causative agent of anthrax) vaccine. After several months of study, however, it became clear that this vaccine showed little efficacy and was discontinued.

Work continued in 1944 at Camp Detrick on a different approach, borrowing from earlier work done in 1905. W. J. Cromartie and D. W. Watson were assigned to the problem, and they found that cell-free extracts from anthrax lesions mimicked the pathology of the disease when injected into animals. Using fluid from anthrax lesions, the task was to isolate the "aggressins" responsible for disease, and then test them as a basis for a vaccine. Significantly, when this "tissue-damaging factor" was repeatedly injected into laboratory animals, it was discovered that rodents thus treated became more and more resistant to anthrax infection, suggesting an immunizing substance was at work. In 1954, investigators found that blood sera from animals infected by *Bacillus anthracis* also possessed a lethal toxin.

The current AVA preparation utilizes the Protective Antigen (PA) obtained from *Bacillus anthracis* culture, and it is the PA component (possibly in addition to other nonspecific types) that provides the main protective, immunizing response. Originally produced by the Michigan Department of Public Health, a privately-held company (BioPort) has since been the manufacturer of anthrax vaccine in the United States, although it has been plagued with delivery and regulatory problems. The AVA is administered at 0, 2, and 4 weeks, then at 6, 12, and 18 months with annual boosters following. Data are scant as to the efficacy in protection against inhalation anthrax in humans, but the AVA vaccine, originally licensed in 1970 and essentially unchanged since its inception, has shown to be very effective in protecting primates from lethal aerosol challenge with anthrax spores. Among the 1590 persons employed at the US Army Medical Research Institute of Infectious Diseases who received altogether 10,451 doses of AVA between 1973 and 1999, 4 percent reported local reaction, while 0.5 percent had systemic reactions, including headache, fever, muscle and joint aches. However, "all local and systemic reactions resolved without any lost time from work or long-term effects."[26]

Because about 150,000 US soldiers were vaccinated during the Persian Gulf War, some have implicated this vaccine as a cause of Gulf War Syndrome (GWS), an illness described generally as a constellation of fatigue, mood-cognition, and musculoskeletal complaints.[27] They have gone so far as to charge a cover-up on the part of the US military, alleging that soldiers were given an "experimental" version of the anthrax vaccine using a different adjuvant, Squalene.[28] (One study found that one arthritis-prone rat strain developed arthritis following an injection of 200 milliliters of Squalene,[29] and others have speculated about the role of diagnostic markers to this and other adjuvants under study.) In an article published in February 2000, investigators purported to have found antibodies to Squalene in self-reporting GWS patients. Unlike previous reports, however, in this research, the scientists make no claim that Squalene was surreptitiously used in the Gulf War anthrax inoculation series.[30] The latter study concludes rather that Squalene antibodies were found in sera drawn from individuals who self-reported GWS symptoms. Recent examinations of randomly selected lots of AVA from 1998 to the present, as well as remaining unopened bottles from previous years, found no evidence of Squalene having been used as an adjuvant.[31]

As for the safety of the anthrax vaccine, over 385,000 personnel had been inoculated as of December 1999, with about 500 reporting adverse reactions. A small percentage of these required hospitalization, several reporting an allergic reaction, while the cause of an adverse reaction for the remainder of the group was not directly associated with the vaccine. No reactions that could be classified as anaphylactic shock, however, had been recorded as of late 1999. Considering the many years of having used the same FDA-approved anthrax vaccine and the large number of people having been vaccinated since the 1950s, criticisms that the vaccine is inherently unsafe and/or ineffective are without foundation.[32]

Today, in addition to receiving routine inoculations for measles, polio, and influenza, depending upon geographical assignment, US soldiers, sailors, and airmen may also be immunized for cholera, Japanese B encephalitis, plague, typhoid, and yellow fever. When taken in the aggregate, the reactivity of the latter vaccines are probably much more pronounced than those found in the administration of AVA. But none of the other required inoculations, including those for influenza and bacterial meningitis, receive nearly as much attention as does AVA.

One writer has suggested that the public's suspicion of the US military, heightened over the past few decades of Agent Orange litigation and the more recent issue of Gulf War Syndrome, is the driving force behind a political and legal movement against the current anthrax vaccine program.[33] It may also be that the DOD has not adequately explained the risks versus benefits of AVA administration to its many charges. (Meant to allay fears among those in the uniformed services, extensive literature on the anthrax vaccine can be found on the Pentagon's website.[34]) But according to the most recent survey (April 12, 2000), conducted by the Tripler Army Medical Center in Hawaii, no detectable "patterns of unexpected local or systemic" reactions to the anthrax vaccine were found among the 425,976 servicemen and women.[35] If administration of AVA is relatively safe, and there does exist, at least according to the Pentagon, a substantial threat from weaponized anthrax, what is the controversy all about?

This is not the first time in history that military personnel have been ordered to be immunized against a possible BW agent, nor is the current policy debate over anthrax vaccinations for the US military unprecedented. Immunization programs to protect soldiers from infectious diseases—whether they be naturally present or in the form of a BW threat—go back at least four centuries, and some of these have also been quite controversial. For example, whether or not to continue vaccination of US troops against smallpox had been hotly debated, especially throughout the 1980s.

The Current Controversy

One viable criticism of the mandatory immunizations could be that because only a small risk of an actual biological weapons attack exists, no matter how safe and effective, the anthrax vaccine is simply not warranted. Here, one needs to rely more upon intelligence gathering and other arcane sources to get a clearer picture of the risks versus benefits. However, considering that the associated health risks of taking the vaccine are extremely low, while the threat of an anthrax weapon being used against the United States is a real possibility, refusing the vaccine presents an unacceptably high danger to unprotected troops and personnel.

An analogous example in the civilian context could be made in the case of polio immunizations, where a 1 in 2.4 million chance exists that an oral polio virus vaccine will actually cause the disease. The risk posed by the vaccine is extremely small, but it is real, and yet poliomyelitis hasn't been seen in the Western Hemisphere since 1991. On an individual basis, a case could be made that the chances of acquiring polio, unless one travels abroad or has contact with people from endemic regions, is even less than that of acquiring vaccine-associated paralytic poliomyelitis. But there is no question that continued polio vaccinations are required in the United States, and indeed a global eradication campaign is underway to completely eliminate the virus.

If the risk of such an attack seems inflated beyond reason, the alternative is to do nothing. But then, if the vaccinations were not carried out, and an anthrax weapon were used against US military personnel, the results would of course be catastrophic. The DOD faced a similarly pressing dilemma during the Gulf War where, as Al Mauroni describes, "If DOD held back on developmental vaccines and pretreatments to troops in the Gulf, and Saddam initiated CB warfare, the outcry would have been deafening."[36]

In the event of a future biological weapon attack, and without adequate protection against anthrax, the immediate families and wider public will watch their men and women dying in large numbers. It is inevitable that the US government, after having already known of the danger for many years, would be excoriated with the pointed question, "Why didn't you do something to protect our soldiers?" The DOD's current mandatory anthrax vaccination program is a rational step toward avoiding such a horrible outcome. Importantly, anthrax vaccinations, in addition to offering some protection against being exposed to anthrax spores, may also serve as a *deterrent* to governments or organizations that might otherwise perpetrate a biological attack against the US military.

Detractors of the mandated inoculations make a number of claims, one that the vaccine has not been proven safe, and, invoking the Nuremberg Code, that using anthrax vaccine to protect against inhalation anthrax is "experimental."[37] These objections are baseless if only because the intent of the vaccine is preventative, but furthermore, in order to produce a study showing efficacy would involve challenging human subjects with deadly anthrax spores, which *is* clearly and precisely unethical (not to mention illegal). On the other hand, the data from non-human primate studies suggest that the current vaccine in use is effective against pulmonary anthrax.

Critics of the DOD's vaccination program have also posited the argument that AVA administration should be made voluntary. But so long as the US military considers anthrax a real threat, by allowing some to refuse what is deemed medically necessary for all would detract from the military's ability to conduct operations, not to mention harming discipline and morale.

We note here that the DOD seems to have implied that Osama bin Laden, the *eminence gris* of a wider terrorist network made up of former and self-styled mujahideen guerilla fighters, possesses anthrax. Note the wording in the Pentagon's response to Congressional criticism of the former's anthrax vaccination program:

> All of our [military] is subject to terrorist attack by anthrax. At least two groups have it. One has tried to use it. It is strategically unwise to wait for an attack before implementing the program.[38]

The reference to the fact that one group "has tried to use it" no doubt points to Aum Shinrikyo's failed attempts in Japan several years ago, but the other group mentioned in the DOD response is difficult to identify with precision. Osama bin Laden's Al Qaeda group has been associated with a desire to obtain biological weapons. Could this be the group to which the DOD is referring? At the same time, it may be a fair question to ask that if the risk to DOD personnel is great enough to warrant their being vaccinated, then is the wider civilian public also entitled to the same protection?

That is a question, unfortunately, that we may have to answer soon.

Notes

CHAPTER 1

1 Wilfred Owen, *The Wilfred Owen Multimedia Digital Archive*, Humanities Computing Unit, Oxford University: http://www.HCU.OX.AC.UK/JTAP

2 John Keegan, *A History of Warfare* (New York: Alfred A. Knopf: 1993).

3 In Keegan's estimate, World War II resulted, all told, in 50 million deaths, much more than than the horrendous total of 20 million suffered in World War I. John Keegan, *A History of Warfare* (New York: Alfred A. Knopf: 1993): p. 50.

4 http://www.nato.int/related/naa/docu/1996/an253stc.htm.

5 It would be a stretch to go from Amiton to VX; it should be noted, however, that Amiton is an extremely toxic insecticide for mammals. Gordon M. Burck, "Chemical Weapons Production Technology and the Conversion of Civilian Production," *Arms Control* (September 1990): p.153.

6 Gradon B. Carter and Graham S. Pearson, "British Biological Warfare and Biological Defense, 1925–45," in Erhard Geissler and John Ellis van Courtland Moon, eds., *Biological and Toxin Weapons: Research, Development and Use from the Middle Ages to 1945*, SIPRI Chemical & Biological Warfare Studies, No. 18 (Oxford: Oxford University Press: 1999): p.183.

7 R. Jeffrey Smith, "Iraq's Drive for a Biological Arsenal; UN Pursuing 25 Germ Warheads It Believes Are Still Loaded With Deadly Toxin," *Washington Post* (November 21, 1997): p.A1.

8 "Brno Lab Faxed Offer of Deadly Bacteria Samples—UK Journalist," Czech News Agency, CTK National News Wire (November 24, 1998).

9 Adapted from David L. Huxsoll, "Rediscovering Biological Warfare," *Chemical & Biological Warfare Proliferation Course* (Washington, DC: Central Intelligence Agency, Biological Warfare Branch: December 1995).

10 Scott Ritter as quoted by Al Venter in an interview for *Jane's Defense Weekly* (October 14, 1998).

11 Karlheinz Lohs, *Synthetic Poisons* (East Berlin: Deutscher Militärverlag: 2nd ed., 1963): p. 91, 192.

12 Difluor is the abbreviation for methylphosphonic difluoride. Didi is a combination of methylphosphonic difluoride and methylphosphonic chloride used by Iraq for their "quick mix" process.

13 *Technologies Underlying Weapons of Mass Destruction*, OTA-BP-ISC US Congress, Office of Technology Assessment (Washington, DC: US Government Printing Office: December 1993): p. 34.

14 Interview, Gennady Lepyoshkin, Monterey, December 1999.

15 Richard O. Spertzel, Robert Wannemacher and Carol Linden, David R. Franz and Gerald W. Parker, "Biological Weapons Proliferation," *Global Proliferation: Dynamics, Acquisition Strategies and Responses, Vol. IV* (Fort Detrick, MD: US Army Medical Research Institute of Infectious Diseases: 1994): p. 20.

16 Interview with William Patrick, November 6, 1998.

17 *The Problem of Chemical and Biological Warfare: Vol. II: CB Weapons Today*, Stockholm International Peace Research Institute (New York: Humanities Press: 1973): p. 72.

CHAPTER 2

1 http://www.state.gov/www/global/arms/treaties/bwc1.html.

2 Amos A. Fries and Clarence J. West, *Chemical Warfare* (New York: McGraw-Hill: 1921): pp. 56, 58, 78, 87–8, 90.

3 Amos A. Fries and Clarence J. West, *Chemical Warfare* (New York: McGraw-Hill: 1921): p. 96.

4 Barton J. Berstein, "Did the US Have More in Store for Japan?" *San Jose Mercury News* (August 4, 1996): p. 36.

5 Victor A. Utgoff, *The Challenge of Chemical Weapons: An American Perspective* (New York: St. Martin's Press: 1991): pp. 21–2, 57.

6 Ad Hoc Committee Report on Chemical, Biological, and Radiological Warfare, June 30, 1950.

7 http://www.mcclellan.army.mil/usacmls/history1.htm.

8 Ed Regis, *The Biology of Doom* (New York: Henry Holt and Company: 1999): p. 210.

9 US Department of Defense, "Continuing Development of Chemical Weapons Capabilities in the USSR," (October 1983): p. 17. Segal puts this number at over 100,000. David Segal, "The Soviet Union's Mighty Chemical Warfare Machine," *Army* 37 (August 1987): p. 27.

10 Aleksandr Solzhenitsyn, *November 1916: The Red Wheel/Knot II*, translated by H. T. Willets (New York: Farrar, Straus and Giroux: 1998): pp. 311–2.

11 Major General N. S. Antonov, *Khimicheskoe Oruzhiye na Rubezhe Dvukh Stoletii* [*Chemical Weapons at the Turn of the Century*] (Moscow: Progress: 1994): pp. 19–28. Text translated by John Hart, Center for Nonproliferation Studies, CBW Nonproliferation Project, 1998.

12 US Department of Defense, *Continuing Development of Chemical Weapons Capabilities in the USSR* (Washington, DC: DOD: 1983): p. 12.

13 Major General N. S. Antonov, *Khimicheskoe Oruzhiye na Rubezhe Dvukh Stoletii* [*Chemical Weapons at the Turn of the Century*] (Moscow: Progress: 1994): pp. 19–28. Text translated by John Hart, Center for Nonproliferation Studies, CBW Nonproliferation Project, 1998.

14 Frederick R. Sidell and David R. Franz, "Overview: Defense Against the Effects of Chemical and Biological Warfare Agents," in Frederick R. Sidell, Ernest T. Takafuji, and David R. Franz, eds., *Textbook of Military Medicine, Part I: Warfare, Weaponry, and the Casualty: Medical Aspects of Chemical and Biological Warfare* (Washington, DC: Borden Institute, Walter Reed Army Medical Center: 1997): p. 3.

15 Monterey-Moscow Study Group on Russian Chemical Disarmament, 1998 report: "*Eliminating a Deadly Legacy of the Cold War: Overcoming Obstacles to Russian Chemical Disarmament*" (Monterey Institute, Center for Nonproliferation Studies: January 1998): http://cns2.miis.edu/pubs/other/mmsg.html.

16 Tim Weiner, "Defector Claims Soviets Had Chemical Warfare Plan," *The New York Times* (February 1998).

17 Milton Leitenberg, "Biological Weapons Arms Control," *Contemporary Security Policy* 17, no. 1 (April 1996): p. 3.

18 Quoted in Milton Leitenberg, "Biological Weapons Arms Control," *Contemporary Security Policy* 17, no. 1 (April 1996): p. 3.

19 Ken Alibek, *Biohazard* (New York, Random House: 1999): p.234.

20 *The CBW Conventions Bulletin*, no. 40 (June 1998): p. 25.

21 Ken Alibek, *Biohazard* (New York, Random House: 1999): p. 112.

22 Quoted in *The CBW Conventions Bulletin*, no. 40 (June 1998): p. 25.

23 US DIA cable, "Iraqi Chemical Munition Characteristics Baseline, Corrected," Filename: 73455727, Pathfinder Record Number: 55727 (September 1990). Former DIA analyst Ken Dombroski brought this to our attention, personal communication, September 8, 1998.

24 Albert J. Mauroni, *Chemical-Biological Defense* (Westport, Connecticut: Praeger: 1998): pp. 28–9.

25 Quoted in Albert J. Mauroni, *Chemical-Biological Defense* (Westport, Connecticut: Praeger: 1998): p. 92.

26 Jeffrey K. Smart, "History of Chemical and Biological Warfare: An American Perspective," in Frederick R. Sidell, Ernest T. Takafuji, and David R. Franz, eds., *Textbook of Military Medicine, Part I: Warfare, Weaponry, and the Casualty: Medical Aspects of Chemical and Biological Warfare* (Washington, DC: Borden Institute, Walter Reed Army Medical Center: 1997): p. 73.

27 "Iraq Responds to UN Requests," *Star Tribune* (Minneapolis, MN: November 28, 1998): p. 14A.

28 This contradicts, however, some US intelligence that, at least from cursory visual evidence, nerve agents or other chemical weapons were not present during the engagement at Al Fao. Apparently, Iraq had amassed significant forces and outnumbered the Iranian defenders to such a degree that chemical weapons may have not been necessary.

29 US Office of the Secretary of Defense, "*Proliferation: Threat and Response*" (Washington, DC: GPO, 1996), accessible via http://www.defenselink.mil/pubs/prolif/toc.html.

30 http://www.stimson.org/cwc/bwprolif.htm#Iran.

31 Ze'ev Schiff, "Interview with Major General Moshe Ya'alon, IDF Intelligence Chief," *Ha'aretz* (May 11, 1997), translated by FBIS.

32 The median lethal dose for most male adults is about 10 milligrams through the skin, or one hundredth of a gram.

33 http://www.stimson.org/cwc/bwprolif.htm#Syria.

34 Fluoroacetate finds use in killing rodent pests and for trapping animals in the fur industry.

35 Danny Shoham, "Chemical and Biological Weapons in Egypt," *The Nonproliferation Review* (Spring–Summer 1998): pp. 48–9, 51.

36 Quoted in Danny Shoham, "Chemical and Biological Weapons in Egypt," *The Nonproliferation Review* (Spring–Summer 1998): p. 54.

37 Robert Waller, "Case Study 2: Libya," *The Deterence Series: Chemical and Biological Weapons and Deterence* (Alexandria, VA: Chemical and Biological Arms Control Institute: 1998): p. 5.

38 http://www.stimson.org/cwc/bwprolif.htm#NorthKorea.

39 Quoted by Joseph S. Bermudez, Jr. in, "Korean People's Army NBC Warfare Capabilities," in US Senate hearings, *Global Spread of Chemical and Biological Weapons* (Washington, DC: GPO, 101st Congress, February–May 1989): p. 553.

40 North Korea Advisory Group, Report to the Speaker, US House of Representatives, November 1999.

41 David E. Sanger, "North Korea Site an A-Bomb Plant, US Agencies Say," *The New York Times* (August 17, 1998).

42 US Office of the Secretary of Defense, "*Proliferation: Threat and Response*" (Washington, DC: GPO: 1996), http://www.defenselink.mil/pubs/prolif/toc.html.

43 Yu Xinhua and Yang Qingzhen, eds., *Shengwu Wuqi Yu Zhanzheng* (Beijing: National Defense Industry Press: 1997): p. viii.

44 US Department of Defense, Office of the Secretary of Defense, "Proliferation: Threat and Response" (2001): p. 15.

45 Chinese publications offer complete reviews of BW agents and their applicability to warfare. For one example, see Yu Shurong and Jin Renjie, eds., *Fangshengwuzhan Yixue* (Shanghai: Shanghai Kexue Jishu Chubanshe: 1986): pp. 6, 11.

46 The other seven are France, United Kingdom, Russia, Japan, India, United States, Taiwan, and Burma as of May 1998. http://www.stimson.org/cwc/declar.htm.

47 US Office of the Secretary of Defense, "*Proliferation: Threat and Response*" (Washington, DC: GPO: 1996), http://www.defenselink.mil/pubs/prolif/toc.html.

48 Methylene deoxymethamphetamine, or ecstasy, an analogue of methamphetamine, is a cross between a stimulant and a hallucinogenic.

49 Suzanne Daley, "South Africa Used Toxins on Enemies, Commission Told," *The New York Times* (June 11, 1998).
50 "Patterns of Global Terrorism—2000," (Washington, DC: Office of the Coordinator for Counterterrorism: June 2000). Available at http://www.state.gov/s/ct/rls/pgtrpt/2000/.
51 "Countering the Changing Threat of International Terrorism," (Washington, DC: National Committee on Terrorism: June 2000). Available at http://www.fas.org/irp/threat/commission.html.

CHAPTER 3

1 Larry C. Johnson, "The Declining Terrorist Threat," *New York Times*, July 10, 2001, p. A23.

2 John V. Parachini, "The World Trade Center Bombers (1993)," in Jonathan B. Tucker, ed., *Toxic Terror* (Cambridge, Massachusetts: MIT Press: 2000): pp. 189, 200.

3 John V. Parachini, "The World Trade Center Bombers (1993)," in Jonathan B. Tucker, ed., *Toxic Terror* (Cambridge, Massachusetts: MIT Press: 2000): p. 202.

4 Benjamin Weiser, "The Trade Center Verdict," *New York Times*, November 13, 1997, pA1.

5 David E. Kaplan, "Aum Shinrikyo (1995)," in Jonathan B. Tucker, ed., *Toxic Terror* (Cambridge, Massachusetts: MIT Press: 2000): p. 218.

6 Personal communication, Dr. Anthony T. Tu, October 1, 2001.

7 Anthony T. Tu and Naohide Inoue, *Kagaku-Seibutsu Heiki Gairon* [*Overall View of Chemical and Biological Weapons*] (Tokyo: Jiho Press: 2000): p. 15.

8 Kevin Flynn, "Oklahoma City Case Still Missing a Link," *Denver Rocky Mountain News*, April 19, 1998, p. A4.

9 William J. Broad and Judith Miller, "Government Report Says 3 Nations Hide Stocks of Smallpox," *New York Times*, June 13, 1999, p. A1.

10 William J. Broad and Judith Miller, "Government Report Says 3 Nations Hide Stocks of Smallpox," *New York Times*, June 13, 1999, p1. Smallpox vaccinations for the US military were mentioned as late as in a 1990 article, but were no doubt superceded by publication date. Ernest T. Takafuji and Philip K. Russell, "Military Immunizations: Past, Present, and Future Prospects," *Infectious Disease Clinics of North America*, March 1990, p. 151.

11 Linnea Capps, Sten H. Vermund, and Christine Johnsen, "Smallpox and Biological Warfare: The Case for Abandoning Vaccination of Military Personnel," *American Journal of Public Health* 76, no. 10, October 1986, p. 1230.

12 Ken Alibek, *Biohazard* (New York: Random House: 1999): p. 112.

13 According to Ken Alibek, "Contagious agents such as smallpox and plague were intended for long-range, strategic attacks against the territories of the United States, Great Britain, and some other European countries, because nobody wanted to use these weapons close to our own troops." Jonathan B. Tucker, "Biological Weapons in the Former Soviet Union: An Interview with Dr. Kenneth Alibek," *The Nonproliferation Review* 6, no. 3, Spring–Summer 1999, p. 2.

14 D. A. Henderson et al., "Smallpox as a Biological Weapon: Medical and Public Health Management," *Journal of the American Medical Association* 281, no. 22, June 9, 1999.

15 Lecture, D. A. Henderson, Conference on Bioterrorism, La Jolla, California, February 2000.

16 Stephen C. Fehr, "Worries About Public Disclosure, Threat of Terrorism," *Washington Post*, October 10, 1999, p. C11.

17 Carl Hulse, "Group Puts Disaster Data on Internet," *New York Times*, September 12, 1999, p. 32.

18 Stephen C. Fehr, "Worries About Public Disclosure, Threat of Terrorism," *Washington Post*, October 10, 1999, p. C11.

19 Pushpa S. Mehta, Anant S. Mehta, Sunder J. Mehta, and Arjun B. Makhijani, "Bhopal Tragedy's Health Effects," *Journal of the American Medical Association* 264, no. 21, December 5, 1990, pp. 2782–3.

20 Pushpa S. Mehta, Anant S. Mehta, Sunder J. Mehta, and Arjun B. Makhijani, "Bhopal Tragedy's Health Effects," *Journal of the American Medical Association* 264, no. 21, December 5, 1990, p. 2781.

21 Some have suggested that phosgene played a role. Considering that the MIC had already been produced long before the incident, and that phosgene was not in use at the time, makes this less likely. However, it is possible that phosgene was present in limited amounts as a decontaminant, or in side reactions with MIC and water.

22 Dan Kurzman, *A Killing Wind* (San Francisco: McGraw-Hill: 1987): p. 106. Just a few months prior to the Bhopal tragedy, Prime Minister Indira Gandhi was assassinated by Sikh bodyguards, and this no doubt provided ready targets for suspicion. The probably imaginary Sikh terrorist group was named "Black September," the same name of a PLO faction that orchestrated the massacre at the Olympic games in Munich. Jackson B. Browning, *Union Carbide: Disaster at Bhopal* (Union Carbide Corporation: 1993).

23 Personal communication, January 17, 2000.

24 Ashok S. Kalekar, "Investigation of Large-Magnitude Incidents: Bhopal as a Case Study," Presented at the Institution of Chemical Engineers Conference on Preventing Major Chemical Accidents, London, England, May 1988.

25 Statement for the Record of Donald L. Dick, Deputy Assistant Director Counter Terrorism Division, and Director, National Infrastructure Protection Center Federal Bureau of Investigation on "Terrorism: Are America's Water Resources and Environment at Risk?" A report read before the House Committee on Transportation and Infrastructure Subcommittee on Water Resources and Environment Washington, DC.

26 Greg Winter and William J. Broad, "Added Security for Dams, Reservoirs and Aqueducts," *The New York Times,*.September 26, 2001.

27 Alex Nussbaum, "Water Utilities Say Supplies Are Safe, Attack Is Unlikely," *The Record*, September 28, 2001.

28 Alex Nussbaum, "Water Utilities Say Supplies Are Safe, Attack Is Unlikely," *The Record*, September 28, 2001.

29 Carolyn Petersen, "Cryptosporidium and the Food Supply," *Lancet* 345, no. 8958, May 6, 1995, pp. 1128–1129.

30 Greg Winter and William J. Broad, "Added Security for Dams, Reservoirs and Aqueducts," *The New York Times*, September 26, 2001.

31 Statement for the Record of Donald L. Dick, Deputy Assistant Director Counter Terrorism Division, and Director, National Infrastructure Protection Center Federal Bureau of Investigation on "Terrorism: Are America's Water Resources and Environment at Risk?" A report read before the House Committee on Transportation and Infrastructure Subcommittee on Water Resources and Environment Washington, DC.

32 Greg Winter and William J. Broad, "Added Security for Dams, Reservoirs and Aqueducts," *The New York Times,*.September 26, 2001.

33 Alex Nussbaum, "Water Utilities Say Supplies Are Safe, Attack Is Unlikely," *The Record*, September 28, 2001.

34 "Whitman Announces Water Protection Task Force," Washington, DC: *US Newswire*, October 5, 2001.

35 Sarah Lyall, "Foot-and-Mouth Flares Again in Britain," *International Herald Tribune Online*, August 6, 2001.

36 US GAO report, *West Nile Virus Outbreak: Lessons for Public Health Preparedness*, GAO/HEHS-00-180 (Washington, DC: September 2000). See also K. E. Steele, M. J. Linn, R. J. Schoepp, *et al.*, "Pathology of Fatal West Nile Virus Infections in Native and Exotic Birds During the 1999 Outbreak in New York City, New York," *Veterinary Pathology* 37, no. 3, May 2000, pp. 208–24.

37 Health care professionals are directed to a number of sources they already have in consensus statements in the *Journal of the American Medical Association*, *New England Journal of Medicine*, and others. In the case of anthrax, for example, one very good article can be found: Faina A. Abramova, Lev M. Grinberg, Olga V. Yampolskaya, and David H. Walker, "Pathology of Inhalational Anthrax in 42 Cases from the Sverdlovsk Outbreak of 1979," *Proceedings of the National Academy of Sciences, USA* 90, March 1993, pp. 2291–4.

CHAPTER 4

1 Quoted in Dennis Avery, "The Fallacy of the Organic Utopia," in Julian Morris and Roger Bate, eds., *Fearing Food: Risk, Health & Environment* (Boston: Butterworth Heinemann: 1999): p. 8.
2 Ronald F. Bellamy and Russ Zajtchuk, eds., *Textbook of Military Medicine, Part I, Vol. 5, Conventional Warfare: Ballistic, Blast and Burn Injuries* (Washington, DC: Walter Reed Army Medical Center: 1990): p. 49. Copper sulfate is no longer recommended for treating wounds contaminated with white phosphorus. Ibid, p. 340.
3 Timothy C. Marrs, Robert L. Maynard, and Frederick R. Sidell, *Chemical Warfare Agents: Toxicology and Treatment* (New York: John Wiley & Sons: 1996): p. 49.
4 Siegfried Franke, *Manual of Military Chemistry, Vol. 1: Chemistry of Chemical Warfare* [*Lehrbuch der Militärchemie der Kampfstoffe*] (East Berlin: Deutscher Militärverlag: 1967): p. 27.
5 Cheng Shuiting and Shi Zhiyuan, *Chemical Weapons* (Beijing: People's Liberation Army Press: 2nd ed. 1999). Reference taken from the 2nd printing (January 2000): p. 43.
6 From US Army Field Manual (FM 3-9), quoted in James A. F. Compton, *Military Chemical and Biological Agents* (Caldwell, New Jersey: The Telford Press: 1987): p. 111.
7 In 1994, the United States produced 12 million tons of chlorine gas. *Chemical & Engineering News* (June 26, 1995).
8 As a Schedule 3 chemical in the CWC listing of prohibited use or stockpiling for war, phosgene "may be produced in large commercial quantities for purposes not prohibited under [the CWC]." http://www.opcw.nl.
9 Ken Baylor, abstract of dissertation, "Biochemical Studies on the Toxicity of Isocyanates," University College Cork, Ireland (May 1996).
10 Ashkok S. Kalelkar, *Investigation of Large-Magnitude Incidents: Bhopal as a Case Study* (Cambridge, Massachusetts: Arthur D. Little, Inc.: 1988), presented at the Institution of Chemical Engineers Conference on Preventing Major Chemical Accidents, London, UK, May 1988. This report makes a very convincing case for sabotage having set off the major release of MIC. While Frederick Sidell has suggested that phosgene also could have played a role in the Bhopal catastrophe, this doesn't appear to have been the case. http://www.bhopal.com/CaseStudy.html.
11 Amos A. Fries and Clarence J. West, *Chemical Warfare* (New York: McGraw-Hill: 1921): p. 17. In order to produce chlorine in large quantities for the Great War, the United States built a complex that included two buildings, each 541 feet long and 82 feet wide. Gigantic reservoirs held a capacity of 4000 tons of salt, supplying the 200 tons of salt used per day for the electrolytic production of chlorine gas. Because salt is essentially one part sodium and one part chlorine, chlorine was split off in the form of a gas through a top vent, while a lower part eliminated caustic soda (sodium-based compound). Using electricity to separate the chloride from salt, each of the 1750 cell tanks could produce 60 pounds of chlorine in 24 hours.
12 John S. Urbanetti, "Toxic Inhalational Injury," in Frederick R. Sidell, Ernest T. Takafuji, and David R. Franz, eds., *Textbook of Military Medicine, Part I: Warfare, Weaponry, and the Casualty: Medical Aspects of Chemical and Biological Warfare* (Washington, DC: Borden Institute, Walter Reed Army Medical Center: 1997): pp. 256–7.
13 Karlheinz Lohs, *Synthetic Poisons* (East Berlin: Deutscher Militärverlag: 2nd ed., 1963): p. 51.
14 Siegfried Franke, *Manual of Military Chemistry, Vol. 1: Chemistry of Chemical Warfare* [*Lehrbuch der Militärchemie der Kampfstoffe*] (East Berlin: Deutscher Militärverlag: 1967): p. 87.
15 In an otherwise excellent and informative booklet, *The Biological & Chemical Warfare Threat* (CIA 1997): p. 26, was the declarative sentence: "Phosgene is more effective than chlorine because it is slowly hydrolyzed by the water in the lining of the lungs, forming hydrochloric acid that rapidly destroys the tissue."

16 Karlheinz Lohs, *Synthetic Poisons* (East Berlin: Deutscher Militärverlag: 2nd ed., 1963): p. 61.

17 Eric R. Taylor, *Lethal Mists* (Commack, New York: Nova Science Publishers, Inc.: 1999): p. 79. Phosgene reacts strongly with those groups found in amino acids, the amine, hydroxy, and sulfhydryl.

18 Frederick R. Sidell, John S. Urbanetti, William J. Smith, and Charles G. Hurst, "Vesicants," in Frederick R. Sidell, Ernest T. Takafuji, and David R. Franz, eds., *Textbook of Military Medicine, Part I: Warfare, Weaponry, and the Casualty: Medical Aspects of Chemical and Biological Warfare* (Washington, DC: Borden Institute, Walter Reed Army Medical Center: 1997): p. 198.

19 "*Possible Long-Term Health Effects of Short-Term Exposure to Chemical Agents. Vol. 2: Cholinesterase Reactivators, Psychochemicals, and Irritants and Vesicants*" (Washington, DC: National Research Council: 1984): p. 105.

20 Frederick R. Sidell, John S. Urbanetti, William J. Smith, and Charles G. Hurst, "Vesicants," in Frederick R. Sidell, Ernest T. Takafuji, and David R. Franz, eds., *Textbook of Military Medicine, Part I: Warfare, Weaponry, and the Casualty: Medical Aspects of Chemical and Biological Warfare* (Washington, DC: Borden Institute, Walter Reed Army Medical Center: 1997): p. 201.

21 Registered trademark of Merck. Used in combination chemotherapeutic regimens, i.e., MOPP: mustine (Mustargen), vincristine (Oncovin), procarbazine, prednisone (not to be confused with other MOPP and NBC protection clothing).

22 Probably from Charles Hederer and Mark Istin, *L'Arme Chimique et ses Blessures* (Paris: J. B. Bailliere et Fils: 1935).

23 James A. F. Compton, *Military Chemical and Biological Agents* (Caldwell, New Jersey: The Telford Press, 1987): p. 63

24 James A. F. Compton, *Military Chemical and Biological Agents* (Caldwell, New Jersey: The Telford Press, 1987): p. 66.

25 Augustin M. Prentiss, *Chemicals in War* (New York: McGraw-Hill: 1937): p. 166.

26 Siegfried Franke, *Manual of Military Chemistry, Vol. 1: Chemistry of Chemical Warfare* [*Lehrbuch der Militärchemie der Kampfstoffe*] (East Berlin: Deutscher Militärverlag: 1967).

27 Karlheinz Lohs, *Synthetic Poisons* (East Berlin: Deutscher Militärverlag: 2nd ed., 1963): pp. 132–3.

28 Karlheinz Lohs, *Synthetic Poisons* (East Berlin: Deutscher Militärverlag: 2nd ed., 1963): pp. 139–41.

29 Arthur D. F. Toy and Edward N. Walsh, *Phosphorus Chemistry in Everyday Living* (Washington, DC: American Chemical Society: 1987): p. 285. Work cited by Schräder, G. et al., can be found in *Die Entwicklung neuer insektizider Phosphorsäureester* (Weinheim/Bergstrasse, Germany: Verlag Chemie GmbH: 1963). Soman was developed by Russian chemists following the advice of German specifications. James A. F. Compton, *Military Chemical and Biological Agents* (Caldwell, New Jersey: The Telford Press, 1987): p. 137.

30 DFP is also known as a cholinesterase inhibitor for glaucoma treatment, Floropryl® Ophthalmic.

31 L. H. Sternbach and S. Kaiser, *Journal of the American Chemical Society* 74 (May 1952): p. 2215.

32 James A. F. Compton, *Military Chemical and Biological Agents* (Caldwell, New Jersey: The Telford Press: 1987): p. 296.

33 G. L. Carefoot and E. R. Sprott, *Famine on the Wind: Plant Diseases and Human History* (London: Angus and Robertson: 1967): pp. 17–8.

34 James A. F. Compton, *Military Chemical and Biological Agents* (Caldwell, New Jersey: The Telford Press: 1987): p. 312. Compton lists 5 micrograms of LSD having the same effect as 15 milligrams of mescaline, while Franke reports that doses of LSD less than 10 micrograms have "no psychotropic effect." Siegfried Franke, *Manual of Military Chemistry, Vol. 1: Chemistry of Chemical Warfare* [*Lehrbuch der Militärchemie der Kampfstoffe*] (East Berlin: Deutscher Militärverlag: 1967): p. 302.

35 William A. Buckingham, Jr., *Operation Ranch Hand: The Air Force and Herbicides in Southeast Asia, 1961–1971* (Washington, DC: Office of Air Force History, United States Air Force: 1982): p. 112.

36 Or "lachrymators."

37 Augustin M. Prentiss, *Chemicals in War* (New York: McGraw-Hill: 1937): pp. 132.

38 Augustin M. Prentiss, *Chemicals in War* (New York: McGraw-Hill: 1937): pp. 139–42.

39 Another source says 1871. Frederick R. Sidell, "Riot Control Agents," in Frederick R. Sidell, Ernest T. Takafuji, and David R. Franz, eds., *Textbook of Military Medicine, Part I: Warfare, Weaponry, and the Casualty: Medical Aspects of Chemical and Biological Warfare* (Washington, DC: Borden Institute, Walter Reed Army Medical Center: 1997): p. 316.

40 A Japanese scholar, Kira Yoshie, in a book published in 1992 reported that it was plausible that the Japanese had the capability to use gas warfare at Wushe. "Taiwan Wenhua Xueyuan Zhuban 'Wushe Shirjian Yankiuhui,'" *Baifenzhibai Taiwanren Guandian*, no. 1868 (2000).

41 *Possible Long-Term Health Effects of Short-Term Exposure to Chemical Agents. Vol. 2: Cholinesterase Reactivators, Psychochemicals, and Irritants and Vesicants* (Washington, DC: National Research Council: 1984): p. 159.

42 Frederick R. Sidell, "Riot Control Agents," in Frederick R. Sidell, Ernest T. Takafuji, and David R. Franz, eds., *Textbook of Military Medicine, Part I: Warfare, Weaponry, and the Casualty: Medical Aspects of Chemical and Biological Warfare* (Washington, DC: Borden Institute, Walter Reed Army Medical Center: 1997): pp. 310, 315. Some have alleged that the cyanide moeity from CS was the cause of death for at least some of the Branch Davidians. The allegation is apparently supported by the characteristic, post-mortem positions of the bodies that one finds in cyanide poisoning. However, the bodies were so badly burned that the cyanide hypothesis may not stand up to the typical, "pugilist" behavior of bodies that curl and bend upon excessive heat. Finally, it is difficult to deliver enough CS at any one point to provide sufficient cyanide to cause fatalities, and indeed, the cyanide was more likely to have evolved from the burning of plastics as it does in residential fires.

43 Air Force Pamphlet 14-210, A4.8.3., USAF Intelligence (February 1, 1998).

44 Taken *verbatim* from "CWC Facts and Fiction," issued by the Office of the President of the United States (April 4, 1997).

45 Department of the Army, *US Army Activity in the US Biological Warfare Programs* I (February 24, 1977): pp. 2–3.

46 Wang Qiang, Yang Qingzhen, eds., *Wuqi Yu Zhanzheng Jishi Congshu: Huaxue Wuqi Yu Zhanzheng* [*Book Series on Weapons and War: Chemical Weapons and Warfare*] #14 (Beijing: Guofang Gongye Chubanshe: 1997): p. 2. I rely upon this source out of necessity, for it is one of the few extant PRC writings on the subject, and it is reasonably competent. While it distorts the historical record with regard to allegations of US having used chemical and biological weapons, especially during the Korean and Vietnamese conflicts, it probably reflects the current wisdom among the CBW cognoscenti in the People's Liberation Army.

47 Curt Wachtel, *Chemical Warfare* (Brooklyn, New York: Chemical Publishing Co., Inc.: 1941): p. 21.

48 Victor Lefebure, *The Riddle of the Rhine: Chemical Strategy in Peace and War* (New York: The Chemical Foundation, Inc.: 1923): p. 218.

49 $CH_3CH_2CH_2CH_2SH$. One may also find this compound listed under other synonyms, including butanethiol, 1-butanethiol, n-butanethiol, 1-mercaptobutane.

50 The resulting compound is off 3-methyl-2-butene-1-thiol. Lee W. Janson, *Brew Chem 101: The Basics of Homebrewing Chemistry* (Pownal, Vermont: Storey Communications, Inc.: 1996): pp. 73–74.

51 James S. Ketchum and Frederick R. Sidell, "Incapacitating Agents," in Frederick R. Sidell, Ernest T. Takafuji, and David R. Franz, eds., *Textbook of Military Medicine, Part I: Warfare, Weaponry, and the Casualty: Medical Aspects of Chemical and Biological Warfare* (Washington, DC: Borden Institute, Walter Reed Army Medical Center: 1997): p. 292.

CHAPTER 5

1 Thucydides, *The Peloponnsian War*, Book II, Chapter 77, , translated by Rev. Henry Dale (London: Henry G. Bohn: 1851): pp. 1, 138–9.

2 Amos A. Fries and Clarence J. West, *Chemical Warfare* (New York: McGraw-Hill: 1921): p. 1.

3 Joseph Needham, *Science and Civilisation in China, Vol. 5, Part 7: Military Technology: The Gunpowder Epic* (New York: Cambridge University Press: 1986): p. 2.

4 Joseph Needham and Robin D. S. Yates, *Science and Civilisation in China, Vol. 5, Part 6, Military Technology: Missiles and Sieges* (Cambridge, UK: Cambridge University Press: 1994): pp. 470–1.

5 ibid.

6 Joseph Needham, *Science and Civilisation in China, Vol. 5, Part 7: Military Technology: The Gunpowder Epic* (New York: Cambridge University Press: 1986): pp. 43, 66. Needham provides the authoritative source on gunpowder and *"meng huo you,"* or the Chinese version of Greek Fire, as well as the appearance of gunpowder in the tenth century.

7 Edward B. Vedder, *The Medical Aspects of Chemical Warfare* (Baltimore: Williams & Wilkins Company: 1925)

8 Literally "fire medicine," the name is derived from its earlier Taoist, alchemical origins.

9 Joseph Needham, *Science and Civilisation in China, Vol. 5, Part 7: Military Technology: The Gunpowder Epic* (New York: Cambridge University Press: 1986): pp. 83–4.

10 The character, rendered as Pin by Needham, is Yun in modern dictionaries. It could be a classical deviation.

11 Joseph Needham, *Science and Civilisation in China, Vol. 5, Part 7: Military Technology: The Gunpowder Epic* (New York: Cambridge University Press: 1986): p. 89.

12 "Pi-li-pao," Chinese onomatopoeia for explosions.

13 *"Zhen Tian lei,"* literally "Quake-thunderclap."

14 Quoted in Tenney L. Davis, *Chemistry of Powder and Explosives* (Las Vegas, Nevada: Angriff Press: reprint facsimile of 1943 ed.): p. 29.

15 Wyndham D. Miles, "Chapters in Chemical Warfare II: The Chemical Shells of Lyon Playfair (1854)," in *Armed Forces Chemical Journal* 11, no. 6 (November–December 1957): p. 23.

16 Wyndham D. Miles, "Chapters in Chemical Warfare II: The Chemical Shells of Lyon Playfair (1854)," *Armed Forces Chemical Journal* 11, no. 6 (November–December 1957): pp. 23, 40.

17 Among other important work in physics and chemistry, Faraday is credited with having originally discovered benzene.

18 Wyndham Miles, "Suffocating Smoke at Petersburg," *Armed Forces Chemical Journal* 13, no. 4 (July–August 1959): p. 35.

19 Ibid.: pp. 26–7.

20 Wyndham D. Miles, "Chemical Warfare in the Civil War," *Armed Forces Chemical Journal* 12, no. 2 (March–April 1958): pp. 26–7.

21 Ethyl chlorohydrin or ethylene chlorohydrin, ethyl chlorhydrin, 2-chloroethanol, etc. It is basically alcohol (ethanol) with a chlorine atom attached, used as an organic solvent and intermediate for dyes, pharmaceuticals, etc.

22 Thiodiglycol, because of its immediacy to the production mustard, is a Schedule 2 precursor in the lists of the Chemical Weapons Convention (CWC).

23 James K. Senior, "The Manufacture of Mustard Gas in World War I [Part I]," *Armed Forces Chemical Journal* 12, no. 5 (September–October 1958): p. 17.

24 Martin Gilbert, *Winston S. Chuchill, Vol. IV* (Boston: Houghton Mifflin Company: 1975): pp. 913–4. This passage has ben quoted elsewhere, including Paul Johnson's excellent *Modern Times* (1992).

25 Martin Gilbert, *Winston S. Chuchill, Vol. IV* (Boston: Houghton Mifflin Company: 1975): pp. 910, 913–4.

26 Donald Richter, *Chemical Soldiers: British Gas Warfare in World War I* (Lawrence, Kansas: University Press of Kansas: 1992): p. 150.

27 Quoted in Donald Richter, *Chemical Soldiers: British Gas Warfare in World War I* (Lawrence, Kansas: University Press of Kansas: 1992): p. 161.

28 Donald Richter, *Chemical Soldiers: British Gas Warfare in World War I* (Lawrence, Kansas: University Press of Kansas: 1992): p. 160.

29 Edward Spiers, *Chemical Warfare* (Hong Kong: The Macmillan Press: 1986): p. 25

30 Chaim Weizmann, *Trial and Error* (New York: Harper & Brothers: 1949): p.172.

31 An industrial chemist, Carl Bosch, who jointly won the Nobel Prise with Friedrich Bergius in high-pressure studies in 1931, later improved upon the Haber method.

32 "Notwithstanding the maxim and the trauma, much of Haber's research and activity after World War I can be considered patriotic ventures. Indeed, his previous great achievement of nitrogen fixation solved a difficulty which had affected Germany more than any other country. In 1913 Germany purchased about 33 percent of the total Chile nitrate production. The second best customer was the United States with 23 percent. With the Haber process perfected, Germany became independent of Chile's nitrate." Morris Goran, "The Present-Day Significance of Fritz Haber," *American Scientist* 35, no. 3 (July 1947): p. 400.

33 Chaim Weizmann, *Trial and Error* (New York: Harper & Brothers: 1949): p. 172.

34 Chaim Weizmann, *Trial and Error* (New York: Harper & Brothers: 1949): pp. 172–4; and footnote in Bernard D. Davis, Renato Dulbecco, Herman N. Eisen et al., *Microbiology* (New York: Harper & Row, 1967): p. 63. Britain issued the Balfour Declaration in 1917: "His Majesty's Government views with favor the establishment in Palestine of a national home for the Jewish people, and will use their best endeavors to facilitate the achievement of this object, it being clearly understood that nothing shall be done which may prejudice the civil and religious rights of existing non-Jewish communities in Palestine or the rights and political status enjoyed by Jews in any other country." http://www.us-israel.org/jsource/History/balfour.html.

35 *Niespulver* was ortho-dianisidine chlorosulphonate. Stockholm International Peace Research Institute (SIPRI), *The Problem of Chemical and Biological Warfare: Vol. I: The Rise of CB Weapons* (New York: Humanities Press: 1971): p. 27.

36 Morris Goran, *The Story of Fritz Haber* (Norman, Oklahoma: University of Oklahoma Press: 1967): p. 82.

37 Tony Ashworth, *Trench Warfare, 1914–1918* (New York: Holmes & Meier Publishers: 1980): p. 1. Quotation from p. 59. Although hand grenades could have been used in these close combat situations, tacit rules developed by both sides ("it would have made life intolerable") throughout World War I disallowed lobbing bombs at one another's trenches. p. 118.

38 Quoted in Karlheinz Lohs, *Synthetic Poisons* (East Berlin: Deutscher Militärverlag: 2nd ed., 1963): p. 51.

39 Amos A. Fries and Clarence J. West, *Chemical Warfare* (New York: McGraw-Hill: 1921): pp. 1, 11.

40 Quoted in Frederick Sidell, Marrs and Maynard, *Chemical Warfare Agents: Toxicology and Treatment* (New York: John Wiley & Sons: 1996): pp. 161–2.

41 Amos A. Fries and Clarence J. West, *Chemical Warfare* (New York: McGraw-Hill: 1921): p. 150.

42 Report from Zanetti, contemporary of Fries. Amos A. Fries and Clarence J. West, *Chemical Warfare* (New York: McGraw-Hill: 1921): p. 151.

43 Victor Lefebure, *The Riddle of the Rhine* (New York: The Chemical Foundation, Inc.: 1923): p. 238.

44 Victor Lefebure, *The Riddle of the Rhine* (New York: The Chemical Foundation, Inc.: 1923): pp. 238, 241.

45 John Keegan, *The First World War* (New York: Vintage Books: 1998).

46 Richard M. Prize, *The Chemical Weapons Taboo* (Ithaca, NY: Cornell University Press: 1997): p. 2.

47 Edward B. Vedder, *The Medical Aspects of Chemical Warfare* (Baltimore: Williams & Wilkins Company: 1925): p. 258.

48 Curt Wachtel, *Chemical Warfare* (Brooklyn, New York: Chemical Publishing Co., Inc.: 1941) footnote 1, pp. 3, 46–7.

49 Russel H. Ewing, "The Legality of Chemical Warfare," *American Law Review* 61 (January–February 1927): p. 59.

50 Richard Pipes, *Russia under the Bolshevik Regime* (New York: Alfred A. Knopf: 1993): p. 388.

51 Richard Pipes, *Russia under the Bolshevik Regime* (New York: Alfred A. Knopf: 1993): p. 386.

52 Nicolas Werth, "From Tambov to the Great Famine," in *The Black Book of Communism* (Cambridge, Massachusetts: Harvard University Press: 1993): p. 117.

53 Nicolas Werth, "From Tambov to the Great Famine," in *The Black Book of Communism* (Cambridge, Massachusetts: Harvard University Press: 1993): p. 117.

54 Nicolas Werth, "From Tambov to the Great Famine," in *The Black Book of Communism* (Cambridge, Massachusetts: Harvard University Press: 1993): p. 117.

55 Paran is from the Taiwanese aboriginal language. Chuang Chi-ting, "Wushe Memories Highlight Modern Dilemmas," *Taipei Times Online* (October 27, 2000).

56 According to a Taiwan aboriginal activist report, some 10,000 Taiwanese indigenous people were killed during this period. Alliance of Taiwan Aborigines, "Report of Alliance of Taiwan Aborigines Presentation to the United Nations Working Group on Indigenous Populations, from 19th to 30th of July [1993]," Center for World Indigenous Studies, Olympia, Washington, http://www.cwis.org.

57 A Japanese scholar, Kira Yoshie, reported in a book published in 1992 that it was plausible that the Japanese had the capability to use gas warfare at Wushe. "Taiwan Wenhua Xueyuan Zhuban 'Wushe shirjian Yankiuhui,'" *Baifenzhibai Taiwanren Guandian*, no. 1868 (2000).

58 Col. Stanley D. Fair, "Mussolini's Chemical War," *Army* (January 1985): pp. 46, 52.

59 Historian John Keegan considers Hemingway's depiction as being "one of the greatest literary evocations of military disaster." John Keegan, *The First World War* (New York: Vintage Books: 1998): p. 349.

60 Col. Stanley D. Fair, "Mussolini's Chemical War," *Army* (January 1985): p. 52.

61 Col. Stanley D. Fair, "Mussolini's Chemical War," *Army* (January 1985): p. 52.

62 Joseph Needham, later author of the *Science and Civilization in China* series, was a British diplomatic officer who was among the first to report these attacks.

63 Victor A. Utgoff, *The Challenge of Chemical Weapons: An American Perspective* (New York: St. Martin's Press: 1991): p. 30.

64 Victor A. Utgoff, *The Challenge of Chemical Weapons: An American Perspective* (New York: St. Martin's Press: 1991): p. 32.

65 Wang Qiang, Yang Qingzhen, eds., *Wuqi Yu Zhanzheng Jishi Congshu: Huaxue Wuqi Yu Zhanzheng* [*Book Series on Weapons and War: Chemical Weapons and Warfare*] #14 (Beijing: Guofang Gongye Chubanshe: 1997): p. 97.

66 Figure is from Yu Zhongzhou in Liu Huaqiu, ed., *Arms Control and Disarmament Handbook* (December 2000): p. 320.

67 Communication from the Chinese Delegation, signed by Hoo Chi-Tsai, No. 170/938, Geneva (September 5, 1938): p. 1.

68 Federation of American Scientists (FAS), "Taiwan, Chemical Weapons," http://www.fas.org/nuke/guide/taiwan/cw/.

69 Wang Qiang, Yang Qingzhen, eds., *Wuqi Yu Zhanzheng Jishi Congshu: Huaxue Wuqi Yu Zhanzheng* [*Book Series on Weapons and War: Chemical Weapons and Warfare*] #14 (Beijing: Guofang Gongye Chubanshe: 1997): p. 101.

70 Said FDR in 1937: "While, unfortunately, the defensive necessities of the United States call for study of the use of chemicals in warfare, I do not want to aggrandize or make permanent any special bureau of the Army or Navy engaged in these studies." Victor A. Utgoff, *The Challenge of Chemical Weapons: An American Perspective* (New York: St. Martin's Press: 1991): p. 20.

71 Quoted in Joachim Krause and Charles K. Mallory, *Chemical Weapons in Soviet Military Doctrine* (San Francisco: Westview Press: 1992): p. 95.

72 http://www.historyplace.com/worldwar2/timeline/v1.htm.

73 Prime Minister's Personal Minute, July 6, 1944, quoted in Barton J. Bernstein, "Why We Didn't Use Poison Gas in World War II," *American Heritage* 36 (August–September 1985): p. 42.

74 Hermann Ochsner, *History of German Chemical Warfare in World War II, Part I: The Military Aspect* (United States: Historical Office of the Chief of the Chemical Corps: 1949): p. 35.

75 Translated by Kathryn Weathersby, "Deceiving the Deceivers: Moscow, Beijing, Pyongyang, and the Allegations of Bacterial Weapons Use in Korea." See also Milton Leitenberg, *New Evidence on the Korean War*, Cold War International History Project (March 1999).

76 Milton Leitenberg, letter to David Ignatius of *The Washington Post*, February 20, 1989.

77 AP, "Pentagon: No Substance to CNN Allegations on Defectors, Nerve Gas," Tuesday, July 21, 1998; 10:59 A.M. EDT.

78 Peter Arnett, *Live from the Battlefield* (New York: Simon & Schuster: 1994): pp. 140–143. General Singlaub has said that incapacitating agents were considered, but only CS was ever approved for the southeast theater of operations. Personal communication, August 13, 1998.

79 There are differing accounts and versions of what exactly happened at Halabja. An ITN documentary film on Iran-Iraq War mentioned cyanide, and at least as far as many of the civilian deaths are concerned, former DIA analyst Ken Dombroski thinks that cyanide was also the major culprit. Personal communication, September 2, 1998.

80 Christine Gosden "Why I Went, What I Saw," *The Washington Post* (March 11, 1998): p. A19.

81 Anthony H. Cordesman and Abraham R. Wagner, *The Lessons of Modern War, Vol. II: The Iran-Iraq War* (Boulder, Colorado: Westview Press/Mansell Publishing Ltd.: 1990): p. 513.

82 "Iran Now Producing Chemical Weapons," *Jane's Defence Weekly* (June 7, 1986).

CHAPTER 6

1 Russel H. Ewing, "The Legality of Chemical Warfare," *The American Law Review* 61 (January–February 1927): p. 63.

2 Author of the highly influential treatise, *The Influence of Sea Power upon History, 1660–1783* (1918).

3 Amos A. Fries and Clarence J. West, *Chemical Warfare* (New York: McGraw-Hill: 1921): p. 6.

4 Emphasis added. Article 23 (a) and (2), Hague Conference of 1899, II Convention, found in Russel H. Ewing, "The Legality of Chemical Warfare," *The American Law Review* 61 (January–February 1927): p. 62.

5 Amos A. Fries and Clarence J. West, *Chemical Warfare* (New York: McGraw-Hill: 1921): p. 6.

6 *Kölnische Zeitung*, June 26, 1915, quoted in http://www.sipri.se/cbw/research/cbw-continuity.html.

7 Russel H. Ewing, "The Legality of Chemical Warfare," *The American Law Review* 61 (January–February 1927): pp. 63, 67.

8 Victor A. Utgoff, *The Challenge of Chemical Weapons: An American Perspective* (New York: St. Martin's Press: 1991): p. 15.

9 They are probably referring to the effects of large volumes of carbon monoxide and other toxic compounds that result from the use explosives.

10 Russel H. Ewing, "The Legality of Chemical Warfare," *The American Law Review* 61 (January–February 1927): pp. 69, 71.

11 The "Protocol for the Prohibition of the Use in War of Asphyxiating, Poisonous or Other Gases and of Bacteriological Methods of Warfare" entered into force on February 8, 1928. Early signatories included the United States, Germany, Iraq, and Russia..

12 Viruses, though not yet detected visually, were known to the scientific community at the time of the Geneva Protocol, and presumably would also have been included.

13 Chemical Weapons Convention, http://www.opcw.nl/.

14 http://www.opcw.nl/.

15 Mustargen by Merck, an alkylating chemotherapy agent, is one of those compounds that has peaceful uses, applied in relatively small amounts for the treatment of cancers.

16 http://www.opcw.nl/guide.htm#chemical.

17 Perfluoroisobutylene [1,1,3,3,3-Pentafluoro-2-(trifluoromethyl)-1-propene] is a fluorinated olefin.

18 Chemical Weapons Convention, http://www.opcw.nl/.

19 http://www.opcw.nl/guide.htm#chemical.

20 Chemical Weapons Convention, http://www.opcw.nl/.

21 USC Title 22—*Foreign Relations and Intercourse.* Chapter 65: "Control and Elimination of Chemical and Biological Weapons." § 5603, US export controls.

22 http://www.opcw.org/ptshome2.htm.

23 Gordon M. Burck, "Chemical Weapons Production Technology and the Conversion of Civilian Production," *Arms Control* (September 1990): p. 134.

24 Chemical Weapons Convention, http://www.opcw.nl/.

25 Related to the author by a US Army veteran in December 1997 who prefers remaining anonymous.

26 Jonathan B. Tucker, "Viewpoint: Converting Former Soviet Chemical Weapons Plants," *The Nonproliferation Review* (Fall 1996): pp. 78, 85.

27 "Science in Russia. The Diamonds in the Rubble," *The Economist* (November 8, 1997): p. 25.

CHAPTER 7

1 Quoted in Stephen Endicott and Edward Hagerman, *The United States and Biological Warfare: Secrets from the Early Cold War and Korea* (Bloomington: Indiana University Press: 1998): p. 63.

2 This is a suggestion by Ken Alibek, November 6, 1998.

3 Adapted in whole and in part from US Congress, Office of Technology Assessment, *Proliferation of Weapons of Mass Destruction: Assessing the Risks*, OTA-ISC-559 (Washington, DC: US Government Printing Office: August 1993): p. 77.

4 This is not always the case. Before 1969, US offensive biological warfare research emphasized the necessity for availability of treatment, whereas, according to Ken Alibek, in the former Soviet Union antibiotic resistance was bred into some pathogens.

5 Richard Preston, "The Bioweaponeers," *The New Yorker* (March 9, 1998): p. 63.

6 Brad Roberts, "New Challenges and New Policy Priorities for the 1990s," in Brad Roberts, ed., *Biological Weapons: Weapons of the Future?* (Washington, DC: Center for Strategic and International Studies: 1993): pp. 75–77.

7 Richard Preston, "The Bioweaponeers," *The New Yorker* (March 9, 1998): p. 60. Concerned at the time that the test was still classified, Bill Patrick was not willing to share with the author of the article exactly what agent was being used. However, he indicated that it was treatable with antibiotics, therefore it was likely a bacteria or rickettsial agent, probably the former.

8 Testimony of General Colin Powell, US Congress, House Committee on Armed Services, *Hearings on National Defense Authorization Act FYI 1994—HR 2401*, 103rd Cong., 1st sess., H201-33 (Washington, DC: Government Printing Office: 1993): p. 112.

9 Quoted in *Proliferation Brief* 1, no. 8 (July 1, 1998), http://www.ceip.org/programs/NPP/brf8.htm.

10 There is some evidence, however, that viruses can grow in cell-free extracts and under the right conditions.

11 Serendipity and not-so-accidental discovery also can play a role in isolating particularly virulent types of microbial agents. For example, in 1956, the former Soviet Union eventually adopted one particular anthrax strain (836) for its offensive biological weapons program. This strain, isolated from a sewer rat in 1956, traces its origins from a leak of anthrax bacteria from a BW-related facility in Kirov during the early 1950s. Ken Alibek, *Biohazard* (New York: Random House: 1999): p. 78. Alibek tells a rather similar story concerning tularemia having been released into the environment in Kirov, Russia, also having infected rodents with a Schu strain obtained from the United States in the 1950s. It is difficult to reconcile these two purported events. See Richard Preston, "The Bioweaponeers," *The New Yorker* (March 9, 1998): pp. 56–7. "The anthrax strain isolated by the scientist Vladimir Sizov from a sewer system of the city, seemingly had several passages through some susceptible animals. Since the accident with spilling the reactor with anthrax occurred in 1953 and the strain was isolated in 1956, I cannot say more than I said in the book. I know the number '836' was given because of the sample number, from which this strain was isolated. I have worked with Vladimir Sizov for years and he has told me a story how he isolated the strain (he received his PhD for this work)." Personal communication, December 20, 1999.

12 Ken Alibek, interview, November 6, 1998. The Soviet defector, Vladimir Pasechnik, was the first to tell the West that the Soviets had developed plague bacteria resistant to antibiotics. See Richard Preston, "The Bioweaponeers," *The New Yorker* (March 9, 1998): p. 58.

13 Richard Preston, "The Bioweaponeers," *The New Yorker* (March 9, 1998): p. 63.

14 Edward M. Eitzen, "Use of Biological Weapons," in Frederick R. Sidell, Ernest T. Takafuji, and David R. Franz, eds., *Textbook of Military Medicine, Part I: Warfare, Weaponry, and the Casualty: Medical Aspects of Chemical and Biological Warfare* (Washington, DC: Borden Institute, Walter Reed Army Medical Center: 1997): p. 442.

15 Lise Wilkinson, *Animals and Disease* (New York: Cambridge University Press: 1992): p. 27.

16 Michael R. Gilchrist, "Disease & Infection in the American Civil War," *The American Biology Teacher* 60, no. 4 (April 1998): p. 258.

17 Caused by streptococcal infection.

18 As the German bacteriologist, Robert Koch, recalled, Carl Weigert "was the first to use aniline staining for demonstrating bacteria in tissues." Herbert A. Lechevalier and Morris Solotorovosky, *Three Centuries of Microbiology* (San Francisco: McGraw-Hill: 1965): pp. 50, 80.

19 As read before the French Academy of Sciences, April 29, 1878. Found at http://www.fordham.edu/halsall/mod/1878pasteur-germ.html.

20 Wolfgang K. Joklik, Hilda P. Willett, and D. Bernard Amos, eds., *Zinsser Microbiology* (Norwalk, Connecticut: Appleton-Century-Crofts: 18th ed. 1984): p. 6.

21 Michael B. A. Oldstone, *Viruses, Plagues, and History* (New York: Oxford University Press: 1998): p. 8.

22 David M. Locke, *Viruses: The Smallest Enemy* (New York: Crown Publishers: 1974): pp. 17–20.

23 Alexander D. Langmuir, "Epidemiology of Airborne Infection," *Bacteriological Reviews* 25, no. 3 (September 1961): p. 178. In 1961, Langmuir wrote, "The field of airborne infection has had not John Snow to lay down sound theoretical principles early in its development." Ibid, p. 173.

24 The infectious arthroconidia of *Coccidioides immitis* range from 3 to 4.5 microns in width, and 3 to 12 microns in length. Stanley C. Deresinski, "Coccidioides immitis," in Sherwood L. Gorbach, John G. Bartlett, and Neil R. Blacklow, eds., *Infectious Diseases* (Philadelphia: W. B. Saunders & Company: 1992): p. 1917.

25 John Samuelson and Franz von Lichtenberg, "Infectious Diseases," in Ramzi S. Cotran, Vinay Kumar, and Stanley L. Robbins, eds., *Robbins Pathologic Basis of Disease* (Philadelphia: W.B. Saunders Company: 5th ed. 1994): p. 328.

26 LeRoy D. Fothergill, "Biological Warfare and Its Defense," *Armed Forces Chemical Journal* 12, no. 5 (September–October 1958): p. 5.

27 Alexander D. Langmuir, "Epidemiology of Airborne Infection," *Ba7cteriological Reviews* 25, no. 3 (September 1961): pp. 176–7.

28 United Nations, *Health Aspects of Chemical and Biological Weapons* (Geneva: World Health Organization: 1970): p. 75; higher end of 25 percent given in *Medical Management of Biological Casualties Handbook* (Fort Detrick, Maryland: US Army Medical Research Institute of Infectious Diseases: 2nd ed. August 1996), http://www.usamriid.army.mil/Content/FMs/medman.

29 Courtesy of Ed Friedlander, MD, http://www.pathguy.com.

30 It is *not* named after Queensland, Australia, where Q-Fever was first identified. Wolfgang K. Joklik, Hilda P. Willett, and D. Bernard Amos, eds., *Zinsser Microbiology* (Norwalk, Connecticut: Appleton-Century-Crofts: 18th ed. 1984): p. 652; and *Medical Management of Biological Casualties Handbook* (Fort Detrick, Maryland: US Army Medical Research Institute of Infectious Diseases: 3rd ed. 1998), http://www.nbcmed.org/SiteContent/MedRef/OnlineRef/FieldManuals/medman/Handbook.htm.

31 Ken Alibek, interview, November 6, 1998. Ken Alibek notes that Q-Fever was eventually replaced in the Soviet arsenal, much to the chagrin of its progenitor, Urakov, who said something to the effect that it "was a real weapon, but nobody takes it seriously anymore." Ken Alibek, *Biohazard* (New York: Random House: 1999): p. 162.

32 Korean hemorrhagic fever with renal syndrome (HFRS).

33 Approximately 3200 UN forces in Korea (1950–1953) came down with Hantavirus infection. http://www.cdc.gov/ncidod/EID/vol3no2/schmaljo.htm.

34 Jonathan F. Smith, Kelly Davis, Mary Kate Hart, George V. Ludwig, David J. McClain, Michael D. Parker, and William D. Pratt, "Viral Encephalitides," in Frederick R. Sidell, Ernest T. Takafuji, and David R. Franz, eds., *Textbook of Military Medicine, Part I: Warfare, Weaponry, and the Casualty: Medical Aspects of Chemical and Biological Warfare* (Washington, DC: Borden Institute, Walter Reed Army Medical Center: 1997): p. 563.

35 Corrie Brown, "Agro-Terrorism: A Cause for Alarm," testimony before the US Senate Subcommittee on Emerging Threats (October 27, 1999): p. 2.

36 "The outbreak of FMD in Taiwan was caused by the introduction of virus through either the smuggling of goods or related agricultural products. As a consequence, the defense against such smuggling is of great importance. . . . It was finally determined by means of analysis in foreign research institute(s) that the FMD outbreak was absolutely the same as that in the Mainland, thus proving that infection was brought into Taiwan from the PRC." Fang Qingquan, "Cong Kodiyi Tan YangZhu Zhengce," *Nongmu Xunkan*, no. 1265 (September 25, 1999): p. 43.

37 Charles J. Stahl, Christopher C. Green, and James B. Farnum, "The Incident at Tuol Chrey: Pathological and Toxicologic Examinations of a Casualty After Chemical Attack," *Journal of Forensic Sciences* 30, no. 2: p. 328.

38 In 1960, some 100,000 turkeys in Great Britain were affected by aflatoxin from groundnut meal. Kenji Uraguchi and Mikio Yamazaki, *Toxicology, Biochemistry and Pathology of Mycotoxins* (New York: John Wiley & Sons: 1978): p. 7.

39 Al J. Venter, "UNSCOM Odyssey: The Search for Saddam's Biological Arsenal," *Jane's Intelligence Review* 10, no. 3 (March 1998).

40 Analysis by David Franz, satellite course "Medical Response to Biological Warfare and Terrorism," hosted by Ted Cieslak, 1998, sponsored by the US Army Medical Research Institute of Infectious Disease (USAMRIID).

41 A 1965 US Army consent form listed five botulinum toxoids for protection against A, B, C, D, and E type botulinum toxins. Frederick R. Sidell, Ernest T. Takafuji, and David R. Franz, eds., *Textbook of Military Medicine, Part I: Warfare, Weaponry, and the Casualty: Medical Aspects of Chemical and Biological Warfare* (Washington, DC: Borden Institute, Walter Reed Army Medical Center: 1997): p. 61.

42 In California, 10 percent of jarred honey contained *C. botulinum* endospores. Jacquelyn G. Black, *Microbiology: Principles and Applications* (New Jersey: Prentice Hall: 2nd ed. 1993): p. 680.

43 Another dinoflagellate, *Pfiesteria piscicida*, has been responsible for large "lesion fish kills" in the United States. C. Gregory Smith, and Stanley I. Music, "Pfiesteria in North Carolina: The Medical Inquiry Continues," *North Carolina Medical Journal* (August 1998), http://www.junkscience.com/news3/music.htm.

CHAPTER 8

1 *Toxon*, the Greek word for the bow, became the root for toxin, derived from the practice of poisoning arrows. Joseph T. Shipley, *Dictionary of Word Origins* (New York: Philos Library: 1965): p. 195.

2 Mark Wheelis, "Biological Warfare Before 1914," in Erhard Geissler and John Ellis van Courtland Moon, eds., *Biological and Toxin Weapons: Research, Development and Use from the Middle Ages to 1945*, SIPRI Chemical & Biological Warfare Studies, No. 18 (Oxford: Oxford University Press: 1999): p. 14.

3 LTC George W. Christopher, USAF, et al. "Biological Warfare: A Historical Perspective," *JAMA* (August 6, 1997): p. 412.

4 Elizabeth A. Fenn, "Biological Warfare, circa 1750," *The New York Times* (April 11, 1998): p. A11.

5 Peter d'Errico, "Jeffrey Amherst and Smallpox Blankets: Lord Jeffrey Amherst's Letters Discussing Germ Warfare Against American Indians," http://www.nativeweb.org/pages.legal.amherst/lord_jeff.html.

6 Jonathan B. Tucker, *Scourge: The Once and Future Threat of Smallpox* (New York: Atlantic Monthly Press: 2001): p. 21.

7 Mark Wheelis, "Biological Warfare Before 1914," in Erhard Geissler and John Ellis van Courtland Moon, eds., *Biological and Toxin Weapons: Research, Development and Use from the Middle Ages to 1945*, SIPRI Chemical & Biological Warfare Studies, No. 18 (Oxford: Oxford University Press: 1999): pp. 28–9.

8 Stanhope Bayne-Jones, *The Evolution of Preventative Medicine in the United States Army, 1607–1939* (Washington, DC: Office of the Surgeon General, Department of the Army: 1968): p. 51.

9 Formerly *Pseudomonas*, renamed in 1992.

10 Harvey J. McGeorge, "The Deadly Mixture: Bugs, Gas, and Terrorists," *NBC Defense & Technology International* 1, no. 2 (May 1986): pp. 56–61.

11 1925 Geneva Protocal text.

12 Mark Wheelis, "Biological Sabotage in World War I," in Erhard Geissler and John Ellis van Courtland Moon, eds., *Biological and Toxin Weapons: Research, Development and Use from the Middle Ages to 1945*, SIPRI Chemical & Biological Warfare Studies, No. 18 (Oxford: Oxford University Press: 1999): pp. 42, 46, 56–58.

13 Jerzy Witt Mierzejewski and John Ellis van Courtland Moon, "Poland and Biological Weapons," in Erhard Geissler and John Ellis van Courtland Moon, eds., *Biological and Toxin Weapons: Research, Development and Use from the Middle Ages to 1945*, SIPRI Chemical & Biological Warfare Studies, No. 18 (Oxford: Oxford University Press: 1999): p. 66.

14 Erhard Geissler, "Biological Warfare Activities in Germany, 1923–45," in Erhard Geissler and John Ellis van Courtland Moon, eds., *Biological and Toxin Weapons: Research, Development and Use from the Middle Ages to 1945*, SIPRI Chemical & Biological Warfare Studies, No. 18 (Oxford: Oxford University Press: 1999): p. 93.

15 Hal Gold, *Unit 731 Testimony* (Tokyo: Yen Books: 1996): pp. 67–82.

16 George W. Christopher, Theodore J. Cieslak, Julie A. Pavlin, and Edward M. Eitzen, Jr., *Biological Warfare: A Historical Perspective* (Fort Detrick, MD: USAMRIID): p. 3. Accessible via http://140.139.42.105/content/BioWarCourse/HX-3/HX-3.html.

17 Liao Yunchang in Liu Huaqiu, ed., *Arms Control and Disarmament Handbook* (Bejiing: National Defense Press: December 2000): p. 368.

18 Hal Gold, *Unit 731 Testimony* (Tokyo: Yen Books, 1996): pp. 86–7, 96–9.

19 John Ellis van Courtland Moon, "US Biological Warfare Planning and Preparedness: The Dilemmas of Policy," in Erhard Geissler and John Ellis van Courtland Moon, eds., *Biological and*

Toxin Weapons: Research, Development and Use from the Middle Ages to 1945, SIPRI Chemical & Biological Warfare Studies, No. 18 (Oxford: Oxford University Press: 1999): p. 217.

20 It is not certain why tetanus was relegated to this group, considering the criteria chosen. Rexmond C. Cochrane, *History of the Chemical Warfare Service in World War II, Vol. II: Biological Warfare Research in the United States* (Fort Detrick, MD: Historical Section, Plans, Training and Intelligence Division, Office of Chief, Chemical Corps: November 1947): passim.

21 Ed Regis, *The Biology of Doom* (New York: Henry Holt and Company: 1999): p. 224.

22 Stephen Endicott and Edward Hagerman, *The United States and Biological Warfare: Secrets from the Early Cold War and Korea* (Bloomington: Indiana University Press: 1998): pp. 40, 48.

23 Quoted in Stephen Endicott and Edward Hagerman, *The United States and Biological Warfare: Secrets from the Early Cold War and Korea* (Bloomington: Indiana University Press: 1998): p. 87.

24 John Ellis van Courtland Moon, "Biological Warfare Allegations: The Korean War Case," Annals of the New York Academy of Sciences 666 (1992): p. 61.

25 John Ellis van Courtland Moon, "Biological Warfare Allegations: The Korean War Case," Annals of the New York Academy of Sciences 666 (1992): p. 61.

26 Yasuro Naito, "Documents Reveal PRC, DPRK Fabrications," *Sankei Shimbun* (January 8, 1998), Morning Edition: p. 1.

27 David Rees, *Korea: The Limited War* (New York: St. Martin's Press: 1964): p. 359.

28 Bruce B. Auster, "Unmasking an Old Lie," *US News and World Report* (November 16, 1998), http://www.usnews.com.

29 Clemens Work, "Coastal Germ Warfare Test Is Blamed for 1950 Death: Family Sues for $11M," *The National Law Journal* 2 (April 21, 1980): p. 3.

30 These figures are different from Alibek (12 million, 2–10 million, resp.). Jacquelyn G. Black, *Microbiology: Principles and Applications* (New Jersey: Prentice Hall: 2nd ed. 1993): p. 650.

31 Ken Alibek, "Behind the Mask: Biological Warfare," *Perspective* IX, no. 1 (September–October 1998).

32 Ken Alibek, "Behind the Mask: Biological Warfare," *Perspective* IX, no. 1 (September–October 1998).

CHAPTER 9

1 Quoted in Nicholas A. Sims, *The Evolution of Biological Disarmament*, Stockholm International Peace Research Institute. (Oxford, UK, Oxford University Press: 2001): p. 195.
2 The Henry L. Stimson Center, "The House of Cards: The Pivotal Importance of a Technically Sound BWC Monitoring Protocol," http://www.stimson.org/cwc/cards.html: 2001: p. 10.
3 ibid. pp. 10–11.
4 Willy Kempel, Ministry of Foreign Affairs, Austria, "The Biological and Toxin Weapons Convention: A Historic and Political Perspective," presented at A Strengthened Biological and Toxin Weapons Convention: Potential Implications for Biotechnology May 28–29, 1998, Institute of Applied Microbiology, University for Agricultural Science, Vienna, Austria.
5 Charles J. Stahl, Christopher C. Green, and James B. Farnum, "The Incident at Tuol Chrey: Pathologica and Toxicologic Examinations of a Casualty after Chemical Attack," *Journal b Forensic Sciences* 30, no. 2, p. 328.
6 The Royal Society, *Scientific Aspects of Control of Biological Weapons* (London: July 1994): p. 9.
7 Mathew Meselson, Jeanne Guillemin, Martin Hugh-Jones, Alexander Langmuir, Ilona Popova, Alexis Shelokov, and Olga Yampolskaya, "The Sverdlovsk Anthrax Outbreak of 1979," *Science* 266, November 18, 1994, pp. 1206–7.
8 Jeanne Guillemin, *Anthrax: The Investigation of a Deadly Outbreak* (Berkeley: University of California Press: 1999): p. 242.
9 Quoted on the *Frontline* website:
http://www.pbs.org/wgbh/pages/frontline/shows/plague/sverdlovsk/meselson.html.
10 The Royal Society, *Scientific Aspects of Control of Biological Weapons* (London: July 1994): p. 10.
11 Willy Kempel, Ministry of Foreign Affairs, Austria, "The Biological and Toxin Weapons Convention: A Historic and Political Perspective," presented at A Strengthened Biological and Toxin Weapons Convention: Potential Implications for Biotechnology May 28–29, 1998, Institute of Applied Microbiology, University for Agricultural Science, Vienna, Austria.
12 The Royal Society, *Scientific Aspects of Control of Biological Weapons* (London: July 1994): p. 10.
13 Jonathan B. Tucker, "Strengthening the Biological Weapons Convention," *Arms Control Today* 25, no. 3, April 1995, p. 10.
14 Nicholas A. Sims, *The Evolution of Biological Disarmament*, SIPRI Chemical & Biological Warfare Studies, No. 19 (Oxford: Oxford University Press: 2001): p. 38.
15 Raymond Zilinskas, "Cuban Allegations of Biological Warfare by the United States: Assessing the Evidence," *Critical Reviews in Microbiology* 25, no. 5, 1999.
16 Nicholas A. Sims, *The Evolution of Biological Disarmament*, SIPRI Chemical & Biological Warfare Studies, No. 19 (Oxford: Oxford University Press: 2001): p. 42.
17 Ibid.
18 Ibid.
19 Ibid., p. 43.
20 Ibid.
21 Milton Leitenberg, personal communication with the author.
22 John Hart, CNS, October 27, 1998.
23 John Hart, CNS, October 27, 1998.
24 "Owners of US Biotechnology Patents: Surging Growth Dominated by US," *Chemical & Engineering News*, October 19, 1998, p. 82.
25 Gillian R. Woollett, "Industry's Role, Concerns, and Interests in the Negotiation of a BWC Compliance Protocol," in *Biological Weapons Proliferation: Reasons for Concern, Courses of Action*, Henry L. Stimson Center Report No. 24, January 1998, p .45.
26 Ibid.

CHAPTER 10

1 Adapted from table in Michael R. Gilchrist, "Disease & Infection in the American Civil War," *The American Biology Teacher* 60, no. 4, April 1998, p. 258.

2 Yu Xinhua, Yang Qingzhen, eds., *Wuqi Yu Zhanzheng Jishi Congshu: Shengwu Wuqi Yu Zhanzheng* [*Weapons and Warfare: Biological Weapons and War*] (Beijing: Guofang Gongye Chubanshe: 1997): pp. 138–9.

3 Nicolau Barquet, and Pere Domingo, "Smallpox: The Triumph over the Most Terrible of the Ministers of Death," *Annals of Internal Medicine* (October 15, 1997): pp. 635–42.

4 B. H. Liddell Hart, *History of the Second World War* (London: Pan Books: 1973): p. 312. I'm indebted to Gavin Cameron for his advice concerning this part of military history.

5 B. H. Liddell Hart, ed., *The Rommel Papers* (New York: Harcourt Brace, Co.: 1953): fn 1, p. 1271.

6 David J. McClain, "Smallpox," in Frederick R. Sidell, Ernest T. Takafuji, and David R. Franz, eds., *Textbook of Military Medicine, Part I: Warfare, Weaponry, and the Casualty: Medical Aspects of Chemical and Biological Warfare* (Washington, DC: Borden Institute, Walter Reed Army Medical Center: 1997): p. 548.

7 Lise Wilkinson, *Animals and Disease* (New York: Cambridge University Press: 1992): p. 36.

8 Robert Reid, *Microbes and Men* (New York: Saturday Review Press, 1975) p. 11.

9 Robert Reid, *Microbes and Men* (New York: Saturday Review Press, 1975) p. 11.

10 Wrote a contemporary: "One of the first to make use of a statistical comparison in the interest of preventive medicine was the American clergyman, Cotton Mather. He reported to the Royal Society, during the severe epidemic of 1721, that more than one in six of all who took the disease in the natural fashion died; but that out of three hundred inoculated [i.e., by variolation], only about one in sixty died." Shryock, quoted in Stanhope Bayne-Jones, *The Evolution of Preventative Medicine in the United States Army, 1607–1939* (Washington, DC: Office of the Surgeon General, Department of the Army: 1968): p. 19.

11 Stanhope Bayne-Jones, *The Evolution of Preventative Medicine in the United States Army, 1607–1939* (Washington, DC: Office of the Surgeon General, Department of the Army: 1968): p. 52.

12 K. Chimin Wong and Wu Lien-Te, *History of Chinese Medicine* (Shanghai: 1936/Taiwan: Southern Materials Center, Inc.: 2nd ed. 1985): p. 216.

13 Allan Chase, *Magic Shots: A Human and Scientific Account of the Long and Continuing Struggle to Eradicate Infectious Disease by Vaccination* (New York: Morrow: 1982): p. 192.

14 Allan Chase, *Magic Shots: A Human and Scientific Account of the Long and Continuing Struggle to Eradicate Infectious Disease by Vaccination* (New York: Morrow: 1982): p. 192.

15 Allan Chase, *Magic Shots: A Human and Scientific Account of the Long and Continuing Struggle to Eradicate Infectious Disease by Vaccination* (New York: Morrow: 1982): p. 192.

16 V. I. Agafonov and R. A. Tararin, "Some Organizational-Tactical Forms and Methods of Anti-Epidemiological Work in Troops of the Stalingrad (Donsk) Front in 1942–43," *Zhurnal Mikrobiologii, Epidemiologii y Immunobiologii* 52, no. 5 (May 1974): p. 7. Translation by Sarka Krcalova.

17 Robert S. Anderson, Ebbe Curtis Hoff, and Phebe M. Hoff, eds., *Preventive Medicine in World War II, Vol. IX: Special Fields* (Washington, DC: Office of the Surgeon General, Department of the Army: 1969).

18 National Academy of Sciences, Committee on Research in the Life Sciences of the Committee on Science and Public Policy, *The Life Sciences; Recent Progress and Application to Human Affairs; The World of Biological Research Requirements for the Future*, 1970.

19 Allan Chase, *Magic Shots: A Human and Scientific Account of the Long and Continuing Struggle to Eradicate Infectious Disease by Vaccination* (New York: Morrow: 1982): p. 192.

20 Albert B. Sabin, "Epidemic Encephalitis in Military Personnel," *Journal of the American Medical Association* 133 (February 1, 1947): p. 290.

21 An official history of the US Army's work in preventing communicable diseases suggests that this vaccine went untried on humans, but this cannot be entirely correct. John Boyd Coates, Jr., Ebbe Curtis Hoff, and Phebe M. Hoff, eds., *Preventive Medicine in World War II, Vol. VI: Communicable Diseases, Malaria* (Washington, DC: Office of the Surgeon General, Department of the Army: 1963): p. 495. At the very least, one could say that the Japanese B encephalitis vaccine had limited clinical or experimental trial data in 1945.

22 Albert B. Sabin, "Epidemic Encephalitis in Military Personnel," *Journal of the American Medical Association* 133 (February 1, 1947): p. 293.

23 Rexmond C. Cochrane, *History of the Chemical Warfare Service in World War II, Vol. II: Biological Warfare Research in the United States* (1947): p. 150.

24 Thomas W. McGovern, and Arthur M. Friedlander, "Plague," in Frederick R. Sidell, Ernest T. Takafuji, and David R. Franz, eds., *Textbook of Military Medicine, Part I: Warfare, Weaponry, and the Casualty: Medical Aspects of Chemical and Biological Warfare* (Washington, DC: Borden Institute, Walter Reed Army Medical Center: 1997): p. 483.

25 "Surveillance for Adverse Events Associated with Anthrax Vaccination—US Department of Defense, 1998–2000," *Morbidity and Mortality Weekly Report* 49, no. 16 (April 28, 2000), http://www.cdc.gov/epo/mmwr/preview/mmwrhtml/mm4916a1.htm.

26 Arthur M. Friedlander, Phillip R. Pittman, and Gerald W. Parker, "Evidence for Safety and Efficacy Against Inhalational Anthrax," *Journal of the American Medical Association* 282, no. 22 (December 8, 1999). Also found at http://www.anthrax.osd.mil/.

27 Fukuda et al., "Chronic Multisymptom Illness Affecting Air Force Veterans of the Gulf War," *Journal of the American Medical Association* 280, no. 11 (September 16, 1998): p. 981–8.

28 Gary Matsumoto, "The Pentagon's Toxic Secret," *Vanity Fair* (May 1999): p. 82–98.

29 J. C. Lorentzen, "Identification of Arthritogenic Adjuvants of Self and Foreign Origin," *Scandinavian Journal of Immunology* 49 (1999): p. 47.

30 Pamela B. Asa, Yan Cao, and Robert F. Garry, "Antibodies to Squalene in Gulf War Syndrome," *Experimental and Molecular Pathology* 68 (February 2000): pp. 55–64.

31 "Squalene Test Findings," conducted by Stanford Research International, Menlo Park, CA. http://www.anthrax.osd.mil/.

32 Following a 1998 review of known clinical trials using anthrax vaccines, authors of an article concluded that "The results of our review show that there are several limits to the knowledge of the effects of anthrax vaccines. There appear to be few comparative studies available. Available studies assess the older generation of vaccines and show several methodological weaknesses such as uncertain case definitions, unclear vaccination schedules and weak experimental design. Despite such weaknesses and the fact that the two studies in this review assess two different types of vaccines (killed and attenuated), *we believe that the data presented in this review show that overall anthrax vaccines are safe and efficacious.*" Vittorio Demicheli, Daniela Rivetti, Jonathan J. Deeks, Tom Jefferson, and Mark Pratt, "The Effectiveness and Safety of Vaccines against Human Anthrax: A Systematic Review," *Vaccine* 16, no. 9/10 (1998): p. 883.

33 James Terry Scott, "Sticking Point: In Defending Its Troops Against Anthrax, The Pentagon Has Injected Distrust Instead," *Washington Post* (January 30, 2000): p. B1.

34 http://www.anthrax.osd.mil/.

35 From "Surveillance for Adverse Events Associated with Anthrax Vaccination—US Department of Defense, 1998–2000," *Morbidity and Mortality Weekly Report* 49, no. 16 (April 28, 2000), http://www.cdc.gov/epo/mmwr/preview/mmwrhtml/mm4916a1.htm: "The findings indicate that rates of local reactions were higher in women than men and that no patterns of unexpected local or systemic adverse events have been identified."

36 Albert J. Mauroni, *Chemical-Biological Defense: US Military Policies and Decisions in the Gulf War* (Westport, Connecticut: Praeger, 1998): p. 56.

37 Representative Christopher Shays, R-Conn, and chairman of the Government Reform National Security Committee, called the anthrax vaccination program "a well-intentioned but overwrought response to the threat of anthrax." Quoted in Andrea Stone, "Anthrax Vaccines Won't Be Stopped," *USA Today* (February 18, 2000): p. 2A.

38 DOD Response to the Staff Report of the House Government Reform's Subcommittee on National Security, Veterans Affairs and International Relations entitled, "The Department of Defense Anthrax Vaccine Immunization Program: Unproven Force Protection," (February 29, 2000): p. 3. Found at http://www.anthrax.osd.mil/SCANNED/ARTICLES/grabedocs/vaccines.htm.

Select Bibliography

Alibek, Ken. 1999. *Biohazard*. New York: Random House.

The Biological & Chemical Warfare Threat. 1997. Washington, DC: US Central Intelligence Agency.

Brown, Fredrick. 1968. *Chemical Warfare: A Study in Restraints*. Princeton, New Jersey: Princeton University Press.

Burck, Gordon M. and Charles C. Flowerree. 1991. *International Handbook on Chemical Weapons Proliferation*. New York: Greenwood Press.

Compton, James A. F. 1987. *Military Chemical and Biological Agents*. Caldwell, New Jersey: The Telford Press.

Cordesman, Anthony H. 1991. *Weapons of Mass Destruction in the Middle East*. London: Brassey's.

Crone, Hugh D. 1992. *Banning Chemical Weapons: The Scientific Background*. Cambridge: Cambridge University Press.

Frazier, Thomas W. and Drew C. Richardson, eds. 1999. *Food and Agricultural Security, Annals of the New York Academy of Sciences*, vol. 894. New York: The New York Academy of Sciences.

Fries, Amos A. and Clarence J. West. 1921. *Chemical Warfare*. New York: McGraw-Hill.

Geissler, Erhard and John Ellis van Courtland Moon, eds. 1999. *Biological and Toxin Weapons: Research, Development and Use from the Middle Ages to 1945*, SIPRI Chemical & Biological Warfare Studies, no. 18. Oxford: Oxford University Press.

Gold, Hal. 1996. *Unit 731 Testimony*. Tokyo: Yen Books.

Harris, Robert. 1982. *A Higher Form of Killing*. New York: Hill and Wang.

Heller, Charles E. 1984. *Chemical Warfare in World War I*. The Leavenworth papers. Washington, DC: Government Printing Office.

Journal of the American Medical Association 278, August 6, 1997.

Kaplan, David E. and Andrew Marshall. 1996. *The Cult at the End of the World*. New York: Crown Publishers.

Krause, Joachim and Charles K. Mallory. 1992. *Chemical Weapons in Soviet Military Doctrine*. San Francisco: Westview Press.

Lefebure, Victor. 1923. *The Riddle of the Rhine*. New York: The Chemical Foundation.

Mangold, Tom and Jeff Goldberg. 1999. *Plague Wars*. New York: St. Martin's Press.

Marrs, Timothy C. , Robert L. Maynard, and Frederick R. Sidell. 1996. *Chemical Warfare Agents: Toxicology and Treatment*. New York: John Wiley & Sons.

Mauroni, Albert J. 1998. *Chemical-Biological Defense*. Westport, Connecticut: Praeger.

Ochsner, Herman. 1949. *History of German Chemical Warfare in World War II, Part I: The Military Aspect*. United States: Historical Office of the Chief of the Chemical Corps.

Prentiss, Augustin M. 1937. *Chemicals in War: A Treatise on Chemical Warfare*. New York: McGraw-Hill.

Regis, Ed. 1999. *The Biology of Doom*. New York: Henry Holt and Company.

Roberts, Brad, ed. 1993. *Biological Weapons: Weapons of the Future?* Washington, DC: Center for Strategic and International Studies.

Scientific Aspects of Control of Biological Weapons. July 1994. London: The Royal Society.

Sidell, Frederick R., Ernest T. Takafuji, and David R. Franz, eds. 1997. *Textbook of Military Medicine, Part I: Warfare, Weaponry, and the Casualty: Medical Aspects of Chemical and Biological Warfare*. Borden Institute, Walter Reed Army Medical Center: Washington, DC.

Somani, Satu M., ed. 1992. *Chemical Warfare Agents*. San Diego: Academic Press.

Spiers, Edward M. 1989. *Chemical Weaponry*. New York: St. Martin's Press.

Stockholm International Peace Research Institute. 1971. *The Problem of Chemical and Biological Warfare, Volume I: The Rise of CB Weapons*. New York: Humanities Press.

Stockholm International Peace Research Institute. 1973. *The Problem of Chemical and Biological Warfare, Volume II: CB Weapons Today*. New York: Humanities Press.

Taylor, Eric R. 1999. *Lethal Mists*. Commack, New York: Nova Science Publishers.

Tucker, Jonathan B., ed. 2000. *Toxic Terror: Assessing Terrorist Use of Chemical and Biological Weapons*. Cambridge, Massachusetts: MIT Press.

US Congress, Office of Technology Assessment. December 1993. *Technologies Underlying Weapons of Mass Destruction*, OTA-BP-ISC. Washington, DC: US Government Printing Office.

US Senate hearings before the 101st Congress. February–March 1989. *Global Spread of Chemical and Biological Weapons*. Washington, DC: Government Printing Office.

Utgoff, Victor A. 1991. *The Challenge of Chemical Weapons: An American Perspective*. New York: St. Martin's Press.

Vedder, Edward B. 1925. *The Medical Aspects of Chemical Warfare*. Baltimore: Williams & Wilkins Company.

Wachtel, Curt. 1941. *Chemical Warfare*. Brooklyn, New York: Chemical Publishing Co.

Wiegele, Thomas C. 1992. *The Clandestine Building of Libya's Chemical Weapons Factory: A Study in International Collusion*. Carbondale: Southern Illinois University Press.

Williams, Peter and David Wallace. 1989. *Unit 731: Japan's Secret Biological Warfare in World War II*. New York: The Free Press.

Wright, Susan. 1990. *Preventing a Biological Arms Race*. Cambridge, Massachusetts: MIT Press.

Zilinskas, Raymond A., ed. 1999. *Biological Warfare: Modern Offense and Defense*. Boulder, Colorado: Lynne Rienner.

Index

A

acetonitrile, 65
acetylcholine, 111–112
acetylcholinesterase (AChE), 111–112
acrolein ("papite"), 120
adamsite: See DM.
Ad Hoc Group of Governmental
    Experts to Identify and Examine
    Potential Verification Measures
    from a Scientific and Technical
    Standpoint (VEREX), 244
Abramova, Dr. Faina, 241
AEF: See American Expeditionary
    Force.
Aeneias Tacticus, 127
aerosols, 202–203
aflatoxin, 214
Agent Blue, 123
Agent Orange, 28, 123. See also herbi-
    cides.
Agent White, 123
agricultural BW agents, 30
agricultural industries, protecting, 83
Alibek, Ken, 34–35
    on Sverdlovsk, 242–243
alimentary toxic aleukia (ATA), 216–217
American Expeditionary Force (AEF),
    21
American First Gas Regiment, 23
American Type Culture Collection, 11
Amiton, 10
amphetamine analogues ("speed"), 116
anthrax, 67–69, 205–206
    military vaccination program,
        261–265
    Soviet weaponization of, 35
    weaponized by the United States, 31
    World War II and, 30
Antonov, N.S., 33
Arab–Israeli War (1973), 33
Argentine hemorrhagic fever virus,
    210–211
arms control:
    BTWC and, 237–248
    CWC and, 169–188

history (brief), 23
"MacDonald Plan," 23
arsenic, 34, 108
arsenic trichloride, 106, 108
arsine (arseniuretted hydrogen), 108
asymmetrical warfare, 63–64
ATA: See alimentary toxic aleukia.
atropine, 113
Aum Shinrikyo, 64–65

B

BA: See bromacetone.
Bacillus anthracis, 11–12, 203
    animal sources, 11
    Iraqi acquisition of, 11
    Sverdlovsk, at, 241
    weapons developed with, World War
        II, 30
Bacillus globigii (BG), 203
Bacillus subtilis, 203
Bacon, Roger, 130
bacteria, 200–201
    defined, 200
    Gram-negative, 200
    Gram-positive, 200
Baghdad: See Iraq.
bartonite: See iodoacetone.
Basson, Wouter, 116
Beijerinck, Martimus Willem, 202
belladonna, 113
benzyl bromide, 119
benzyl iodide ("fraissite"), 120
"Big Eye," 28
binary chemical weapons, 28
bioaerosols, 202–203
Biological and Toxin Weapons
    Convention, 1972 (BTWC),
        237–248
    articles of, 238–240
    confidence-building measures
        (CBMs), 243
    current problems, 248
    Depositary Governments and, 239
    destruction of weapons, 239

(continued)

differences from the CWC, 237, 246, 248

dual-use issues and, 246

duration of, 239

Middle East and participation in, 20

Nixon and, 31

non-proliferation in, 239

peaceful uses of biological weapons in, 238, 239

protocols for, 240

ratification by the United States, 31

Review Conferences, 240, 243, 244

Russian admission of violation, 34

signing and ratifying, 230

Soviet Union and its alleged violation of, 241, 242. See also Sverdlovsk.

UN and, 239

US and its alleged violation of, 244–245

biological weapons and agents, 6, 193–218

acquisition of, 11

agricultural agents, 30

Bacillus anthracis, 11

ballistic missiles and, 35

"boomerang" effect of, 194

"chimera" virus, 197

chemical weapons, in contrast with, 196

definition of, 6, 193–196

delivery of, 16–18, 35

dual use and production of, 12

effectiveness of, 7

"footprint" of, 197

history of, 219–236

inhalation of, 196

Johnston Atoll tests, 195

Japanese development and use of, 224–225

production of, 13–14

proliferation of, 12, 19–60, 237–247

simulants, 31

terrorism and, 18

types of, 196–218

US development of, 226–229

viral versus bacterial agents, 196

See also specific biological agents.

Biopreparat, 34

Alibek, Ken, 34

biotechnology industry, 245–246, 248

concerns about the BTWC, 246

Pharmaceutical Research and Manufacturers of America (PhRMA), 246

bioterrorism:

recent trends in, 58–61, 63–67

blister agents (vesicants), 98–105

blood agents, 105–109

Bn-Stoff: See brommethylethyl ketone.

Bolivian hemorrhagic fever virus, 210–211

bomomartonite: See brommethylethyl ketone.

Bonjean's Ergotin, 115

botulism: See also botulinum toxin, Clostridium botulinum, 214–215

botulinum toxin, 204

bromacetone ("B-Stoff"), 119

brommethylethyl ketone ("Bn-Stoff"), 119

bromobenzyl cyanide (CA or "camite"), 119, 120

Brucella melitensis

Iraqi acquisition of, 11

US weaponization of, 31

brucellosis: See also Brucella melitensis.

BTWC: See Biological and Toxin Weapons Convention.

Burkholderia mallei, 207

BZ (3-quinuclidinyl benzilate), 24, 113

C

CA: See bromobenzyl cyanide.

camite: See bromobenzyl cyanide.

Camp Detrick, Maryland. See Fort Detrick, Maryland.

Canada:

BW program, United States and, 30
capsaicin, 117
carbon monoxide, 108
Carroll, James, 202
CBW: *See* chemical and biological warfare.
CDC: *See* Centers for Disease Control and Prevention.
Centers for Disease Control and Prevention (CDC), 72–73
Central Scientific Research Military–Technical Institute (TsNIVTI), 33
chemical and biological warfare (CBW) armaments:
    acquiring, 9–13
    advantages of, 7
    defined, 6
    delivery of, 16–18
    fear of, 3, 5
    implied threat of, 9
    production of, 12–14
    proliferators, 9
    utility of, 7
    weaponization of, 14–16
chemical weapons and agents, 87–126
    aniline dye and, 134
    chemical industry in Germany and, 133–136
    coal-tar and, 133–134
    definition of, 6, 87–89
    delivery of, 90–93
    dismantling, 33
    dye industry in Germany and, 10
    history of, 127–168
    indigo dye and, 134
    organic chemistry and, 133–136
    properties of, 89–90
    types of, 92–123
    See also specific chemical agents.
Chemical Corps, US, 26. *See also* US Army, Chemical Corps.
Chemical Manufacturers Association (CMA), 246
Chemical Weapons Convention (CWC),
    10, 19, 28, 175–181
    Australia Group and, 180–181
    challenge inspections and, 175, 184–185
    clean-up of weapons and, 176
    declaration of facilities in, 179
    destruction of weapons and, 176, 186–188
    facilities conversion and destruction in, 188–190
    intrusive verification and, 175
    Middle East and participation in, 20
    monitoring agents and precursors in, 182–183
    precursors and, 177
    proliferation "signatures" and, 183
    Russian chemical dismantlement and, 33–34
    Soviet Union and, 175–176
    schedules of agents and precursors, 177–179
    US and, 21, 28, 172–176
    US export controls in, 180
    verification and, 181–182
China, 54–56
    the BTWC and, 239, 245
Chlamydia psittaci, 203
chloracetone, 119
chloracetophenone (CN), 117, 121
chlorine, 93–95
chloropicrin (chlorpicrin), 96–97, 106
choking agents, 92–95
cholera, 208–209
Churchill, Winston, 136–137, 156–157
CI: *See Coccidioides immitis.*
CK: *See* cyanogen chloride.
Clarke, Hans T., 135
Clostridium botulinum, 215
    Iraqi acquisition of, 11
Clostridium perfringens
    Iraqi acquisition of, 11
CN, (chloracetophenone), 117, 121
Coccidioides immitis (CI), 203
Congo-Crimean hemorrhagic fever virus, 210–211

Convention on the Prohibition of the
    Development, Production, and
    Stockpiling of Bacteriological
    (Biological) and Toxin Weapons
    Convention and Their
    Destruction, the, 240. *See also* the
    BTWC.
conventional weapons, 65–66
Coxiella burnetii, 31, 201. 208. *See also*
    Q–fever.
Compton, James , 113
CR, 117
Crimean War, 131
Crusades, 127–128
CS, 117, 121
Cuba, 58
cults, 7
CWC: *See* Chemical Weapons
    Convention.
CX: *See* phosgene oxime.
cyanogen chloride (CK), 108
cyclite: *See* benzyl bromide.

D
DA: *See* diphenylchlorarsine.
Dengue hemorrhagic fever virus,
    210–211
Department of Defense: *See* US
    Department of Defense (DOD).
DF: *See* methylphosphonic difluoride.
DFP: *See* diisopropylflurophosphate3.
diacetoxyscirpenol (DAS), 214
difluor: *See* methylphosphonic difluo-
    ride.
diisopropylflurophosphate3 (DFP), 111
diphenylchlorarsine (DA), 105
diphosgene, 96
disarmament:
    and biological weapons, 169–192
    and chemical weapons, 237–247
DM ("adamsite"), 122
DOD: *See* US Department of Defense.
dual-use precursors in CW, 10
dual-use technologies in BW, 12

DT, 26
Dugway Proving Grounds, Utah, 26
"*Dulce et Decorum Est*" (Owen), 3

E
Ebola virus, 204, 210–211
ecstasy (drug), 116
Edgewood Arsenal, Maryland, 24, 26
Egypt, 33, 46–47
    CWC and, 20
Ehrlich, Paul, 199
Eighteen-Nation Disarmament
    Committee, 238
ergotin, 115
Ethiopia, chemical attacks on, 152–153
ethylbromacetate, 118
ethyldichlorarsine, 97
ethyliodoacetate (SK), 120

F
Faraday, Michael, 131
Flammenwerfer (German flame-thrower),
    137
fluoroacetates, 80
foot–and–mouth disease (FMD), 30, 83,
    212
Fort Detrick, Maryland, 31
Fracastoro, Girolamo, 198
fraissite: *See* benzyl iodide.
Francisella tularensis, 207
    Iraqi acquisition of, 11
Franco-Prussian War, 131–132
Fries, Amos, 173–174
fungi, 214

G
GA: *See* tabun.
gallium-arsenide, 34
gas gangrene, 12. *See also Clostridium per-
    fringens.*
Gas Service (US), 21
GB: *See* sarin.

GD: *See* soman.
Geneva Protocol ("1925"), 30, 173–175,
    237, 239
    biological weapons and, 223
germ theory of disease, 198
Germany, 33
    CW development in cooperation
        with Soviet Union, 33
Ghosh, Ranajit, 26, 33, 111
glanders, 207–208
Gorbachev, Mikhail, 34
Gram, Hans Christian, 200
Greek fire, 128
G-series nerve agents, 109–111
Gulf War (1991), 28, 166–167
    BTWC, effect on, 243
    CBW, role in, 9
Guillemin, Jeanne, 242
Gunpowder Classic (Chinese Epic), 127, 129

H
Haber, Fritz, 136
Hague Conferences, The, 170
    International Peace Conference of
        1899 (First Hague Conference),
        170
    Second Hague Conference (1907),
        170
Halabja (Iraq), 106, 164
Hama (Syria), 106
Hantavirus, 210–211
harassing agents: *See* Riot Control
    Agents (RCAs).
Hazmat licenses, 75–76
heating, ventilation, and air conditioning
    systems: *See* HVAC.
hemorrhagic fever viruses, 210–211
herbicides, 123–124
Hess, Rudolf, 106
Hitler, Adolf, 23, 33, 159
HN-2, 102
HN-3, 102
Horn Island, Mississippi, 30
HVAC systems, 78

hydrogen cyanide, 105–106
hydrogen sulfide ("sour gas"), 10, 109

I
IG Farben (Germany), 109
"immune" buildings, 78
Imperial Chemical Industries (UK), 26
incapacitants, 112–116
insecticides, 10, 24
insects as disease vectors, 30
International Science and Technology
    Center, Moscow (ISTC), 35
Iran, 42–43, 164
    CBW weapons, motivations to
        acquire, 9
Iran-Iraq War, 162–164
Iraq, 11, 36–42
    acquisition of biological weapons
        and, 11–12
    American Type Culture Collection
        and, 11–12
    biological weapons in, 40
    chemical weapons in, 36
    Gulf War and, 36–37
    VX in, 39
Isopropyl alcohol:
    binary chemical weapons, role in, 28
Israel, 49–50
    civil defense and, 18
    CWC and, 20
Ivanovski, Dimitrii, 202

J
Japan, 26
    Aum Shinrikyo in, 64–65
    plague infestation of Manchuria, 225
    Unit 731, 224

K
Kaffa, siege of (1346), 219–220
Keegan, John, 5
Kircher, Athanasius, 198

Koch, Robert, 199
Korean War, 24–25, 31, 160–161
  allegations of BW, 31

L

lacrimators, 121
Lassa fever virus, 210–211
League of Nations, 23, 136
Lebanon, 20
Lewis, W. Lee, 103
Lewisite, 103–104
Libya, 48–49
  CWC and, 20
lipopolysaccharide (LPS), 200
Livens Projector, 137–140
Livens, William Howard, 137–139
LSD (lysergic acid diethylamide), 115

M

"MacDonald Plan," 23
Mace, 117, 121
Mahan, Alfred T.,
Mahley, Donald, 248
malodorous concoctions, 125
manguinite, 108
Marburg virus, 204, 210–211
Marshall, George C. (Gen.), 23, 155
martonite: See bromacetone.
masking agents, 125
mescaline, 116
Meselson, Mathew, 242–243
Merck, George W., 30
methaqualone (quaaludes), 116
Methylphosphonic difluoride, 10, 28
Meyer, Victor, 135
MMDA, 116
Mo Zi, 127
Münsterlager experimental facility, 106
Muscle Shoals, Alabama, 24
mustard (sulfur), 98–102
  discovery of, 135
  early production of, 136
  precursors for, 10

mycotoxins, 213–214

N

napalm, 124–125
National Pharmaceutical Stockpile
  (NPS), 73
nerve agents, 10, 24, 109–112
  precursors for, 10
  sarin, 24, 65
  soman, 109
  tabun, 109
  VX, 24
Nettle gas: See phosgene oxime.
Niespulver (German sneezing powder),
  141
nitrogen mustard, 102
Nixon, Richard, 21, 26, 35
  BW development, end of, and, 238
  BW policy and, 31
  Soviet perceptions of Nixon's renun-
    ciation of BW, 35
North Korea, 50–53
November 1916 (Solzhenitsyn), 32
NPS: See National Pharmaceutical
  Stockpile.

O

Okinawa, 26
Organization for the Prohibition of
  Chemical Weapons (OPCW), 237
organophosphates, 109
organophosphorous, 24
Owen, Wilfred, 3
  "Dulce et Decorum Est," 3

P

papite: See acrolein.
paraquat, 123
Pasechnik, Vladimir, 34
Pasteur, Louis, 199
Patrick, William III, 34
Peloponnesian War, 127

perfluoroisobutylene (PFIB), 97–98
Perkin, William Henry, 134
Pershing, John J., 172
Petrov, Stanislov (Col. Gen.), 34
peyote cactus (*Lophophora williamsii*), 116
phenyldichlorarsine (PD), 104–105
phosgene, 95–96
phosgene oxime ("nettle gas"),
   103–104
phosphorus pentasulfide, 10
phosphorus trichloride, 10
Pine Bluff, Arkansas, 31
plague, 35, 206–207. *See also Yersinia
   pestis*
   Soviet weaponization of, 35
Playfair, Sir Lyon, 131
"Poor man's atomic bomb," 9
Porton Down (UK), 26
Powell, Colin, 195
precursors, chemical:
   trade in, 10
Prime Time Live, 35
proliferation, 9
protozoa, 217–218
Prussic acid, 106

Q

quaternary ammonium compounds
   ("quats"), 102
Q–fever, 31, 208
QL, 28

R

Rafsanjani, President Ali Akbar
   Hashemi, 9
RCAs: *See* riot-control agents.
Reagan, Ronald, 28
Red Army, Soviet, 33
   Central Military Chemical Proving
      Grounds (TsVKhP), 33
   Central Scientific Research
      Military–Technical Institute
      (TsNIVTI), 33
   Military Chemical School, 33

Reed, Walter, 202
ricin, 216
Ricketts, Harold T., 200
rickettsiae, 200–201
Rift Valley fever, 210–211
riot-control agents (RCAs), 116–122
   banned, 122–123
risk assessment, 18
Rocky Mountain Arsenal, Colorado, 24
Roosevelt, Franklin D.
   BW policy, World War II, 30
   CW policy, World War II, 23
Rickettsia prowazekii: *See* typhus.
Russia:
   "Comprehensive Destruction Act,"
      33
   CW policy, modern, 33
   World War I and CW, 23
   See also Soviet Union.

S

sarin (GB), 26, 65, 109
Saunders, Bernard, 109
saxitoxin, 216
Schräder, Gerhard, 109
Schwartz, Berthold, 130
SEB toxin:
   weaponized by United States, 31
Shahabad (Iran), 106
Shikhany (Russia), 33
Shoah, 105
Sibert, William L., 21
siege warfare, 127
simulants, biological, 31
SK: *See* ethyliodoacetate.
Skull Valley, Utah, 26
smallpox, 30, 70–72, 209–210
   Soviet weaponization of, 35
   weapon against Native Americans, as,
      222
smokes (obscurants), 124
sodium cyanide, 10
Solzhenitsyn, Alexander, 32
soman (GD), 109

South Africa, 57–58
South Korea, 50–54
Soviet Union:
    anthrax and, 241–243
    Biopreparat, 34
    BW program, 20, 34–35
    BTWC and, 20
    Central Scientific Research
        Military–Technical Institute
        (TsNIVTI), 33
    chemical weapons stockpile, 21
    CW program, history, 31–35
    doctrine, 32
    Germany, CW cooperation with, 33
    GRU, 26
    perceptions of Nixon's renunciation
        of BW, 35
    Red Army and CW, 32
    smallpox and, 71
    tabun and, 24
    World War I, CW and, 32
    World War II, 24
    See also Russia.
Stalin, Joseph, 33
stannous chloride, 106
Staphylococcus aureus, 215
staphylococcal enterotoxin B, 215
sternutators, 121
Stevenson Report, 24
Stimson, Henry L., 30, 155
Stokes mortar, 139, 140
sulfur, 28
sulfur mustard (HD), 102
Sverdlovsk, 241–243
    compound at, 19, 241
Syria, 9, 43–45
    CBW weapons, motivations to
        acquire, 9
    CWC and, 20

T

T2: *See* trichloromethane.
tabun (GA), 26, 109
Taiwan, 56–57

Tambov (Russia):
    1921 chemical attack at, 151
tear gas: *See* Riot Control Agents.
    (RCAs).
terrorism.
    CBW and, 8, 20
    recent trends in, 58–61, 63–64
thiodiglycol, 10
thrips palmi, 244–245
Thucydides, 127
TMA (Trimethoxyphenylamino-
    propane), 116
toxins:
    defined, 213
    CWC and, 213
    therapeutic uses of, 213
    See also specific toxins.
Treaty of Versailles (1919), 171–172
trichloromethane,(T2), 106, 214,
    216–217, 241
Truman, Harry S., 23
TsNIVTI: *See* Red Army, Central
    Scientific Research
    Military–Technical Institute.
T-Stoff: *See* xylyl bromide *and* benzyl
    bromide.
TsVKhP: *See* Red Army, Central
    Military Chemical Proving
    Grounds.
Tukhachevsky, Mikhail Mikolaevich, 151
tularemia, 30, 207. *See also Francisella*
    tularensis.
typhus, 30, 201. *See also Rickettsia*
    prowazekii.

U

United Arab Emirates, 20
United Kingdom, 11
    anthrax cakes, production of, World
        War II, 11
    Imperial Chemical Industries and, 26
    response to Soviet BW program rev-
        elations and, 34
    VX and, 26

United Nations Special Commission:
    See UNSCOM.
United States:
    action in World War I, 23
    American First Gas Regiment, 23
    Arab–Israeli War (1973), impact on
        CW policy, 33
    binary chemical weapons and, 28
    biological weapons program, World
        War II, 30
    BTWC and, 21, 31
    BW activity, 30–31
    chemical casualties in World War I,
        23
    Chemical Warfare Service, 21
    chemical weapons, history, 21–28
    chemical weapons stockpile, 21, 24,
        28
    CW agents, 24
    CWC and, 21, 28
    CW policy and, 26, 33
    military and CW defense, 23, 28
    preparedness for bioterrorism and, 18
    revival of chemical weapons pro-
        gram, 28
    sheep kill incident, 26
    Stevenson Report, 24
    war in the Pacific, 23
    War Research Service, 30
    World War I and, 21, 23
    See also entries beginning with "US."
UNSCOM, 13
USAMRIID: *See* US Army Research
        Institute for Infectious Diseases.
US Air Force, 31
US Army, 31
    Chemical Corps, 24
    Chemical Warfare Service, 21
    Research Institute for Chemcal
        Defense, 28
US Army Research Institute for
        Infectious Diseases (USAMRIID),
        73–74
US Clean Air Act of 1990, 75
US Civil War, 131–132

US Environmental Protection Agency,
        75
US Navy:
    "Big Eye"VX program, 28
US public health system, 72–75

V

vaccinations:
    anthrax, 67–69
    history, 250
    military, in, 256–265
    smallpox, 70–72
    typhus, 255–256
van Leeuwenhoek, Antony, 198
Variola major, 210
VE, 111
VEE: *See* Venezuelan equine encephalitis
Venezuelan equine encephalitis (VEE),
        211–212
    weaponized by United States, 31
vesicants: *See* blister agents.
VG, 111
Vietnam War, 26, 161–162
    Chemical Corps activities during, 26
    Stevenson Report, 24
V–gaz, 33
viruses
    defined, 201
    size of, 201–202
    discovery of, 201–202
    tobacco mosaic disease and, 202
vincennite, 106
VM, 111
vomiting agents, 122
von Baeyer, Adolf, 134
VS, 111
V-series nerve agents, 111
VX, 26, 73
    binary form, 28
    espionage, and, 26, 33
    precursors for, 10

W

Washington Arms Conference of 1922, 172–173
War Research Service, United States, 30
water supplies, protecting, 79–82
weaponization process, 14–16
    acquisition process in CBW and, 9–14
    biological, 16–18
    chemical, 14–16, 17
weapons of mass destruction, xvii
Weizmann, Chaim, 140–141
World War I:
    American First Gas Regiment, 23
    anthrax as weapon in, 222
    chemical warfare and, 5
    chlorine, French use of, 145
    discarded munitions, 146
    German reluctance to use biological agents in, 222
    glanders as weapon in, 222
    mustard, first use in, 144, 146
    onset of, 142
    Russian chemical casualties in, 23
    Russian CW capabilities during, 32
    trench construction, 142
    US chemical casualties in, 23
    US participation and, 21
    Ypres, 1915 chlorine attack at, 143–144
World War II:
    Bari incident, 159
    biological weapons in, US, 30
    CW and American opinion, 23
    D-Day Invasion, 155
    Japan and CW, 23
    Japanese munitions in, 154–155
    Japanese mustard attack at Shanxi province, 154
    Marshall, George C., 23
    Pacific campaign, 23
    Roosevelt and CW policy, 23, 154
    Stillwell, Joseph, 23
    V weapons and, 159

Wushe (Taiwan), 1930 chemical attack at, 152

X

xylyl bromide ("T-Stoff"), 119

Y

Yellow fever virus, 30, 210–211
Yeltsin, Boris, 33–34
Yemen:
    1963 war with Egypt, 161
Yersinia pestis, 206
Yousef, Ramsi, 64, 71

Z

Zilinskas, Raymond, 244
Zyklon B, 106

273800

BLAYNEY		
CANOW.		
COWRA		
FORBES	7/02	
MANILDRA		
ORANGE		
MOLONG